TABLES FOR THE REDUCTION OF HINDU DATES

By the same author:

On Hindu Chronology, Acta Orientalia 1922-1926
De Gregoriaansche Kalender, Maastricht 1932
Le Nombre d'Or, The Hague 1936

CHEVALIER JEAN BAPTISTE FRANÇOIS DE WARREN
(JOHN WARREN)

Born at Livorno (Leghorn), September 21, 1769
Died at Pondichéry, February 9, 1830

FOUNDER OF HINDU CHRONOLOGICAL RESEARCH

Reproduced by kind permission after a painting in oil in the possession of Comte Reginald de Warren of Grasse (France)

DECIMAL TABLES

FOR THE

REDUCTION OF HINDU DATES

FROM THE DATA

OF THE

SŪRYA-SIDDHĀNTA

BY

W. E. VAN WIJK

SPRINGER-SCIENCE+BUSINESS MEDIA, B.V.
1938

Originally published by Martinus Nijhoff, The Hague, Holland in 1938
Softcover reprint of the hardcover 1st edition 1938

ISBN 978-94-017-6710-1 *ISBN 978-94-017-6784-2 (eBook)*
DOI 10.1007/978-94-017-6784-2

If it be considered that the doctrines on which these humble Kalendars are calculated, have from time immemorial ruled the Chronology of many civilized and wealthy nations, the subject may not be deemed undeserving of the attention of the votaries of science.

JOHN WARREN

This little book is intended to be useful to epigraphists and interesting to students of technical chronology. I have spared no pains in endeavouring to render the Explanation as intelligible and concise as the subject would allow, and I advise readers not to try to make use of my Tables without having thoroughly studied it.

If the demand for this work proves sufficient I intend to publish a second part dealing with *yogas*, *nakṣatras*, Jovian cycles and reduction to other *Siddhāntas*.

For the mathematical foundations of the Tables I refer to my articles on Hindu Chronology in the Acta Orientalia of the years 1921—26. All calculations have been effected to at least five significant figures; I am indebted to the Dutch Oriental Society for a subvention which enabled me to have part of the work done by others under my supervision. The trouble which my young friends H. W. VERHEYEN, astronomical computor, and A. KUIPERS have taken over the calculatory work and the diagram illustrating the Explanation deserves full appreciation.

My special thanks are due to the good friends who rendered publication possible, to Dr. JOHAN VAN MANEN, secretary to the Oriental Society of Bengal, and to Mr. J. G. BOTH, for procuring me the fine collection of Indian *pañcāṅgas* which forms the foundation of my investigations on the subject: and, not least, to my friend ALEXANDER STOLS, who has again enhanced his printing fame by the fine execution of this small but complicated piece of expert workmanship.

W. E. VAN WIJK

CONTENTS

FOREWORD
BOOKS AND ARTICLES CONSULTED 1
EXPLANATION
Introductory 3
Solar reckoning 4
Lunisolar reckoning, Mean System 6
Lunisolar reckoning, True System 10
The Auxiliary Tables 22
PRACTICAL EXERCICES 26
INDEX 27
THE PROBLEMS ANSWERED 33
TABLES

BOOKS AND ARTICLES CONSULTED

WALTHERUS, THEODORUS, Doctrina temporum indica cum paralipomenis,
EULER, LEONARD, De Indorum anno solari astronomico,
forming the appendix to Th. S. Bayerus, Historia regni Graecorum Bactriani... Petropoli 1738 in 4°.

GATTERER, IOHANNES CHRISTOPHORUS, Chronologia Brahmanvm, forming the preface to I. G. Frank, Novvm Systema Chronologiae fvndamentalis, Goettingae, 1778 in fol.

MARSDEN, WILLIAM, On the Chronology of the Hindoos. Phil. Trans. Vol. *80*, read June 24, 1790. 4°.

JONES, W., On the Chronology of the Hindus. Diss. a. minor pieces rel. to History of Asia, I, 9, 1792 with Supplement, ibid. I, 10 1792 in 8°.

WARREN, JOHN, KALA SANKALITA, a collection of memoirs on the various modes according to which the nations of the southern parts of India divide time... Madras, 1825 in 4°.

SÛRYA SIDDHÂNTA, translation by Rev. Ebenezer Burgess, formerly missionary of the A.B.C.F.M. in India, with notes (by William D. Whitney) and an appendix Journ. Am. Or. Soc., New Haven 1860. *)

BIOT, J. B., On the translation of the Súrya-Siddhánta, Journ. d. Sav. 1860 in 4°.

SÚRYA SIDDHÁNTA, translation by Pundit Ba'pu' Deva Sa'stri, Calcutta, 1861 in 8°.

SPOTTISWOODE, WILLIAM, On the Súrya-Siddhánta and the Hindu method of calculating eclipses. Journ. R. As. Soc. 1863 in 8°.

ALBÊRÛNÎ, 'ABÛ-ALRAIHÂN MUHAMMAD IBN 'AHMAD, India, an accurate description of all categories of Hindu thought, English by E. C. Sachau, Trübner's Oriental Series, London, 1888, 2 voll. in 8°.

DELBOS, LÉON, L'Astronomie aux Indes Orientales, Bull. Sc. math. 1893.

JACOBI, HERMANN, The computation of Hindu dates in inscriptions, etc. Epigraphia Indica 1892.

JACOBI, HERMANN, Tables for calculating Hindu dates in true local time, ibid. 1894.

SEWELL, ROBERT and ŚANKARA BÂLKRISHNA DÎKSHIT, The Indian Calendar with Tables for the conversion of Hindu and Muhammadan into A. D. dates and vice versa. With Tables of eclipses visible in India by Robert Schram, London 1896 in 4°.

SEWELL, ROBERT, Continuation of the „Indian Calendar", Eclipses of the Moon in India, London 1898 in 4°.

THIBAUT, G., Astronomie, Astrologie und Mathematik; Grundriss d. Ind. Ar. Phil. u. Altertumskunde begr. v. G. Bühler, III, 9, 1899.

VELANDAI GOPALA AIYER, The Chronology of ancient India, 1-st and 2-nd series, Madras 1901 sm. in 8°.

SRI KALINATH MUKHERJI, Popular Hindu Astronomy, Part I, Taramandalas and Nakshatras, Calcutta 1905 sm. in 8°.

BARHASPATYAH (pseud. f. Lála Chhota Lál), The obscure text of the Jyotisha Vedanga explained. Allahabad 1907 in 4°.

SCHRAM, ROBERT, Kalendariographische und Chronologische Tafeln, Leipzig 1908.

DEWAN BAHADUR L. D. SWAMIKANNU PILLAI, Indian Chronology, solar, lunar and planetary. A practical guide, Madras 1911. With eye-table in plano.

SEWELL, ROBERT, Indian Chronography, an extension of the „Indian Calendar" with working examples, London 1912 in 4°.

DEWAN BAHADUR L. D. SWAMIKANNU PILLAI, An Indian Ephemeris, A. D. 1800 to A. D. 2000 showing... the ending moments of tithis and nakshatras, Raijapuram, Madras, 1915 in 4°.

*) A new edition, ed. by Phanindralal Ganguly with an introduction by Prabodchandra Sengupta and published by the Calcutta University has appeared in 1935.

VENKATASUBBIAH, A., Some Śaka dates in inscriptions, a contribution to Indian chronology, Mysore 1918.

KIRFEL, W., Die Kosmographie der Inder nach den Quellen dargestellt, Bonn und Leipzig, 1920 in 4°.

VENKATESH BAPUJI KETKAR, Indian and foreign Chronology, Journ. Royal Asiatic Soc., Bombay branch, Bombay and London 1923.

SEWELL, ROBERT, The Siddhantas and the Indian Calendar being a continuation of the author's „Indian Chronography" with an article by the late J. F. Fleet on the mean place of the planet Saturn. Reprinted from Epigraphia Indica, Calcutta 1924, 4°.

DEWAN BAHADUR L. D. SWAMIKANNU PILLAI, Comprehensive Tables for Indian chronology condensed from the author's larger work „Indian Ephemeris" A.D. 700 to A.D. 2000. Madras 1924 fol.

JYOTIS CHANDRA GHATAK, The Conception of the Indian astronomers concerning the precession of the equinoxes, Journ. a. Proc. As. Soc. of Bengal *19*, Calcutta 1924.

KAYE, G. R., Hindu Astronomy, Mem. Arch. Survey of India, Calcutta 1924 in 4°.

SŪRYASIDDHĀNTA, Sanskrit text. ed. by Mahāmahopādhyāya Sudhākara Dvivedī and provided with a commentary called Sudhāvarṣini, 2-nd ed., Bibl. Indica 1925.

VENCATASUBBARAMIAH, C., Handbook of Astrology, s.d.n.l. (Madras 1926).

VARĀHA MIHIRA, The Panchasiddhāntikā, Text, original commentary in Sanskrit, English translation and introduction by G. Thibaut and Mahāmahopādhyāya Sudhākara Dvivedī, Benares 1889. Reprinted by Motilal Banarsi Dass, Lahore 1930.

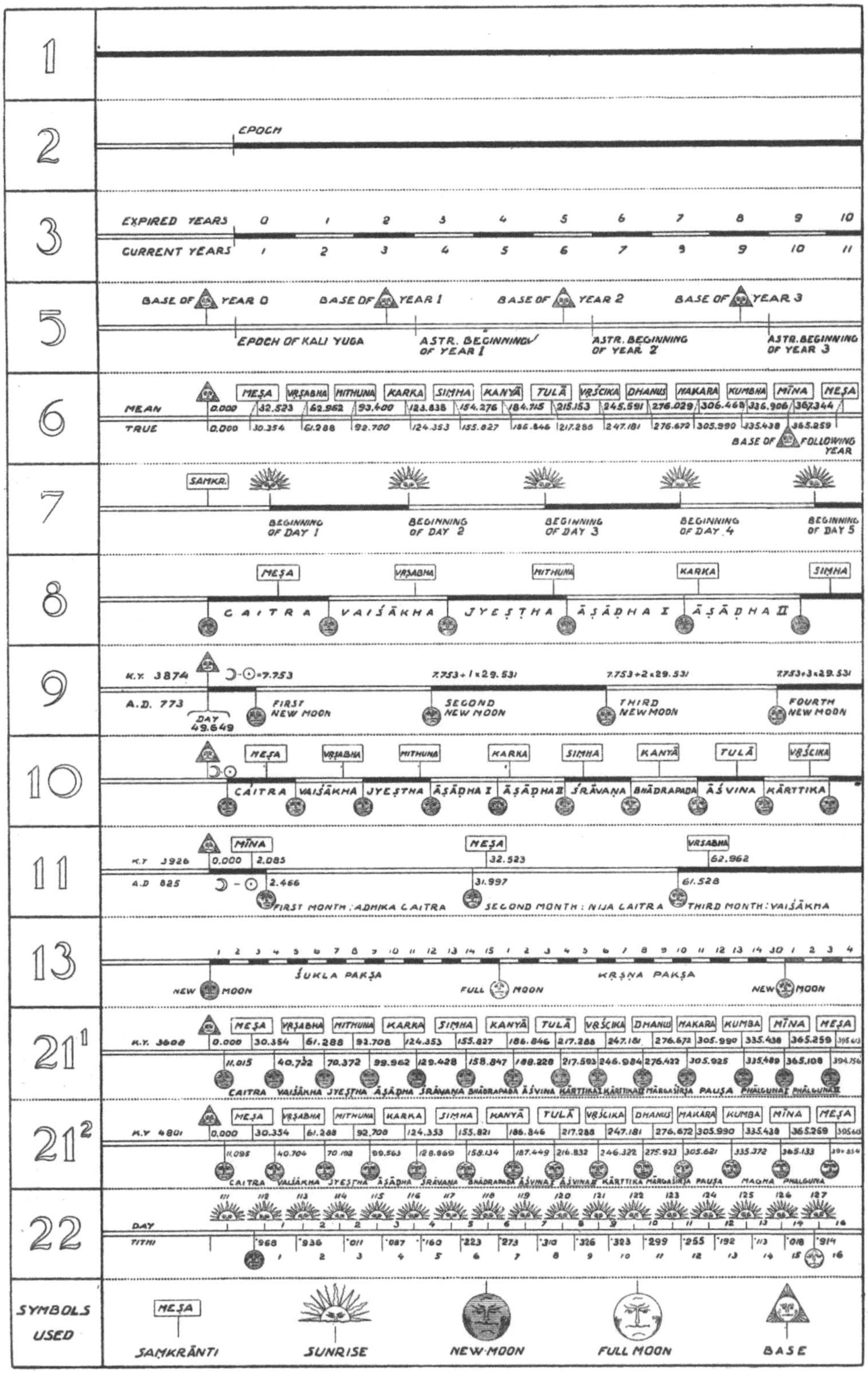

The numbers in the first column refer to the paragraphs of the Explanation.

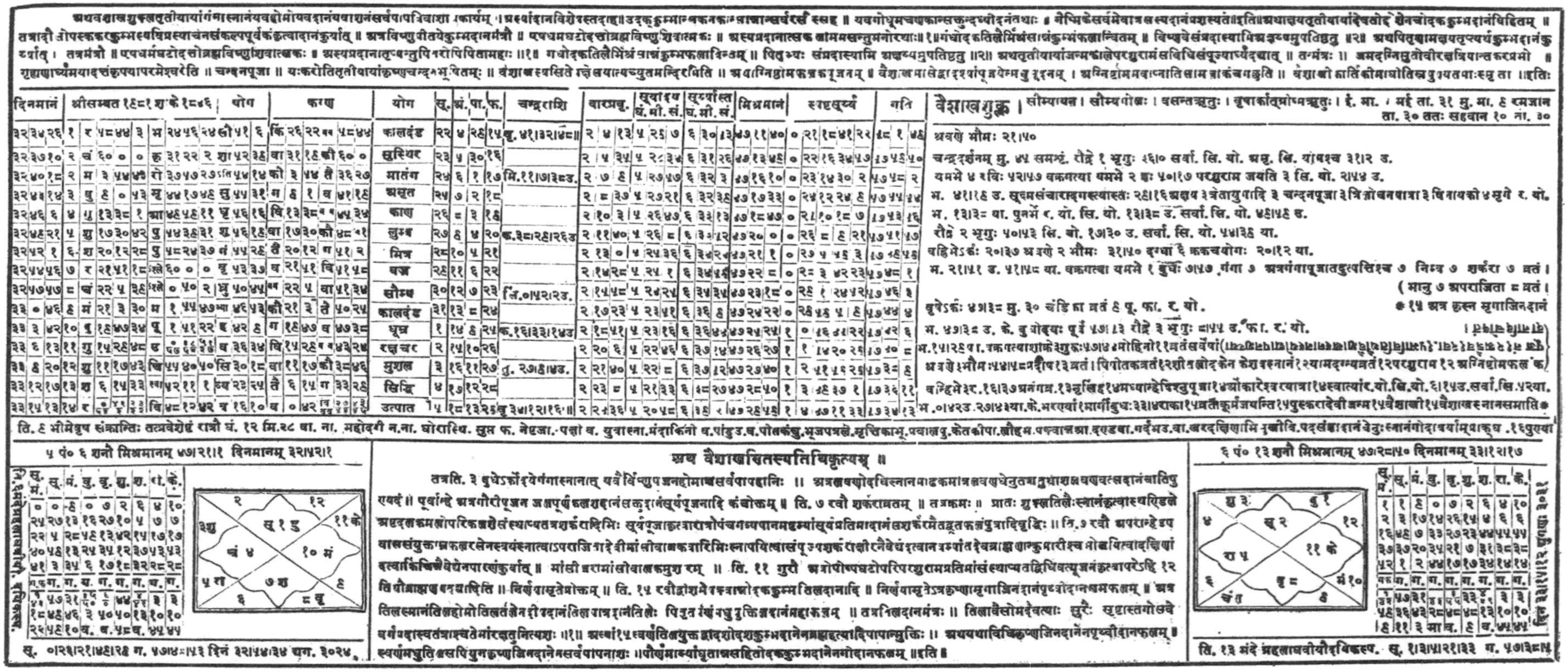

Page from an actual *pañcāṅga* calculated by Shiva Shanker Panday in Rajasthani language in Sanskritic script, showing the bright half of *pūrṇimānta Vaiśākha* of the year 1981 of the *Vikrama* Era, *Śaka* 1846 (=1924/5 A.D.). The second *tithi* is repeated, the 15-th suppressed. The upper part of the page contains prescriptions as to bathing and offerings: „On the third day of the white Moon in the month of *Vaiśākha* one should plunge in the holy Ganges and offer the prescribed things as sacrificial fee . . . . etc." In the middle part the first column gives the duration of the solar day, in *ghaṭikās*, *palas* and *vipalas*, the second column, headed *Samvat* 1981, *Śaka* 1846, contains the ending moments of the *tithis*, the third those of the *yogas*, the following resp. the *karaṇas*, *rāśi*, the moments of sunrise, sunset, meridian passage of the sun and its „motion" (minutes and seconds of arc travalled in 1 day); the last column denotes the hours for performing specific rites and offerings on the preceding *tithis*. The lower part of the page shows the celestial figures for two moments of the month, which are useful in casting horoscopes. The text in the middle contains more sanitary rules and prescriptions concerning offerings and rites.

EXPLANATION

§ 1. TIME. At first sight Hindu chronology seems an intricate matter to the European mind. To explain in a simple way what is necessary for understanding and dealing with the following tables the graphical method seemed to me most expedient. We shall represent TIME by a straight line, without beginning or end. Any inch of that line may stand for a day as well as for a thousand years, for a second as well as for an aeon.

§ 2. EPOCH. Time is measured by man in units comprehensible to the human mind, as days, months and years. Chronology arises when a point of that line is accepted as a starting point to count from; such a starting point is called an EPOCH and the years counted from that epoch form an ERA.

§ 3. EXPIRED AND CURRENT YEARS. The years of an era may be counted in two different ways: the year beginning at the epoch may be considered as year 0 or as year 1 of the era. Both systems are in use in Hindu as in other chronology. The Hindus call the years counted in the first way expired (*gata*) years, in the second way current (*vartamāna*) years.

ILLUSTRATION: We count the years of human life in expired years. A child of seven years has already lived for more than seven years; but on the famous 18 *Brumaire de l'An VIII de la République Française une et indivisible* only 7 years and 47 days of the French Era had elapsed.

Our tables are constructed primarily for expired years of the astronomical era used by the *Sūrya Siddhānta*, called the *Kali Yuga.*

§ 4. EPOCH OF THE *KALI YUGA*. The *Sūrya Siddhānta* accepts $365^{\text{d}}.258756\dot{4}8\dot{1}$ for the astronomical duration of the year. Many different eras are in use, the one with the remotest epoch and therefore embracing all others being the Kali Yuga. The epoch of the *Kali Yuga* coincides with midnight between the 17-th and 18-th day of February of the year 3102 B C (= year -3101 in astronomical reckoning) for the meridian of *Laṅkā.* In these tables the days are assumed to begin at mean sunrise, assumed to be 6 a.m. mean *Laṅkā* time; therefore $48^{\text{d}}.75$ of the year -3101 had elapsed at the moment when the *Kali Yuga* began.

NOTE: The astronomical year of the Hindus is a sidereal year; modern authors on Hindu chronology call it an anomalistic year, but the anomalistic year — according to the *Sūrya Siddhānta* — measures $0^{\text{d}}.0000327211$ more than the sidereal.
The tropical year, which is the astronomical foundation of the Christian era, measures $365^{\text{d}}.242546$.
The civil year, which always counts a whole number of days, can be a good deal longer or shorter than the astronomical year, as will become clear in the course of this explanation.
Laṅkā is a fictitious place on the equator, on the meridian of *Ujjayinī*, the *Avantī* mentioned in the *Sūrya Siddhānta* (I, 62); its longitude is 75°46′6″ East from Greenwich.

§ 5. BASE. For practical reasons these tables are not based on the epoch of the *Kali Yuga* itself but on a moment which precedes it by $32^d5234665$ .. By successively adding 365^d258 .. we get a series of points on the „time-line", each preceding the astronomical beginning of a year of the *Kali Yuga* by 32^d523. These moments we shall call the BASES of the years. It is easy to find the equivalents of these bases in the Julian calendar. The first of them is day 48.750 — 32.523 = day 16.227 of the year -3101; the second is day 16.227 + 365.259 — 365*) = 16.227 + 0.259 of the year -3101 + 1 = -3100; the third 16.227 + 2 × 365.259 — 365 — 366*) = 16.227 + 2 × 0.259 — 1, of the year -3101 + 2, etc.**) To prevent the subtraction of a unit each year after a bissextile the tables accept 15.722 instead of 16.227 as starting point which compels us to increase the numbers for the odd years in column B of Table II by 1. Therefore Table I must always be used in conjunction with Table II.

EXAMPLE: Required the base for the years K.Y. exp. 5000 and 5001.

	A	B	A	B
Table I	5000	59.009	5000	59.009
Table II	00 +	1.000 +	01 +	1.259 +
	5000	60.009	5001	60.268
	3101 —		3101 —	
	A.D. 1899		A.D. 1900	

NOTE: Our BASE is the moment of the true *Mīna saṃkrānti*, which is the nearest moment always to precede the beginning of the *Caitrādi* Hindu civil year. It is chosen with the aim of keeping all calculations with these tables additive on principle.

SOLAR RECKONING

§ 6. *SAṂKRĀNTIS*. Two different forms of year are in use among the Hindus, the first based only on the movement of the sun, the second taking also the moon into account. I shall deal first (in this paragraph and the next) with the solar year.

The Hindu zodiac is divided into 12 signs or *rāśis* and the moment in which the sun in its yearly course enters one of these *rāśis* is called a *saṃkrānti*. A solar year is the time elapsing between two consecutive moments in which the sun enters the same sign; in most cases the *Meṣa saṃkrānti* is considered the astronomical beginning of the year, and such a year is called a *Meṣādi* year. But *Siṃhādi* and *Kanyādi* years also occur.

Before about 4000 K.Y. the *saṃkrāntis* were placed in equal distances on the time-line (therefore each $^1/_{12}$th of a sideral year = 30^d438 removed

*) The year -3101 must be considered a common year, -3100 a leap year, etc.

**) Reference to Tables I and II, columns A and B, will give the Julian equivalent of the base of any year of the Kali Yuga, calculated in this way.

from the next) but afterwards increased knowledge of the astronomical phenomena enabled the calendar-makers to calculate the exact time which the sun needs to proceed 30° in longitude in its course. The distances of these MEAN and TRUE *saṃkrāntis* from the base are given in Section A of Table III; e.g. the mean *Dhanus saṃkrānti* falls $276^{\text{d}}.029$ after the base, the true $276^{\text{d}}.672$, etc.

It is now also possible to find the equivalent of a *saṃkrānti* in the Julian calendar. E.g. we found that the base for the year K.Y. exp. 5001 corresponds to day 60.268 of A.D. 1900; therefore the true *Dhanus saṃkrānti* of that year falls on day $60.268 + 276.672 = 336.940$ of A.D. 1900.

If we wish to know the corresponding date, we have to use Section E of Table III; the year 1900 being a leap year in the Julian calendar, we find $336 - 335 = 1$ December 1900, $0^{\text{d}}.940$ after mean sunrise at *Laṅkā*.

If the Gregorian equivalent is wanted we have — according to Section F of Table III — to add 13^{d}, finding, therefore, December 14 A.D. 1900.

§ 7. SOLAR MONTHS. The solar year is divided into 12 solar months, which receive their names from the *saṃkrāntis*, or from the lunar months which end after these *saṃkrāntis*. The names of these lunar months are also to be found in Section A of Table III. In most cases the first day of the solar month begins at the sunrise next following the *saṃkrānti*.

For other rules for the first day of the solar month see Section E of the first auxiliary table.

EXAMPLE: Required the Julian equivalent for 24 Karka K.Y. exp. 4372, true system.

	A		B
Table I	4300	. . . .	52.879
Table II	72 +	. . . .	1.630 +
	4372		54.509
	3101 −	Table III, true Karka	124.353 +
A.D.	1271		178.862,

which implies that day 1 begins at sunrise of day 179 and day 24 at sunrise of day $179 + 23 = 202$. The year 1271 being a common year, this number — according to Sect. E of Table III — corresponds to $202 - 181 = 21$ July.

LUNISOLAR RECKONING

§ 8. LUNISOLAR YEAR AND MONTHS. The second year form is the lunisolar, and is based on the movements of the moon as well as of the sun. The lunisolar year consists of lunar months or lunations, a lunar month being the time elapsing between two consecutive moments of New Moon. The mean duration of the lunar month is called the synodic period of the moon; according to the *Sūrya Siddhānta* it amounts to $29^{\text{d}}.5305879$.. In most cases the lunation which ends first after the *Meṣa saṃkrānti* is considered the first of the lunar months of the year; this lunation is called *Caitra.*

Again there are two sytems of lunisolar reckoning: the lunations may be considered as having all the same duration, viz. that of the synodic period, or they may be taken as actual intervals between consecutive moments of true conjunctions of sun and moon. The first system, using mean (*madhyama*) lunations is the oldest; the true (*spaṣṭa*) system became prevalent roughly about 4000 K.Y. We have to deal with the mean system first, as the true system presupposes a thorough knowledge of the mean reckoning.

The names of the lunisolar months are given in Section A of Table III.

LUNISOLAR RECKONING. MEAN SYSTEM

§ 9. DISTANCE OF MEAN NEW MOONS FROM BASE. If the distance of the first New Moon from the base is known for a year, all the other New Moons of that year are equally known, as they follow each other at a distance of $29^{d}531$. The distance of the first New Moon from the base is found by means of columns C of Tables I and II, whilst the multiples of the synodic period are given in Section G of Table III.

As the distance of the first New Moon from the base always must be less than $29^{d}531$, that number must be subtracted from the sum of the numbers in columns C of the Tables I and II as soon as this sum exceeds that number. For this reason it appears for convenience sake over column C in Table II.

EXAMPLE: Required the Julian equivalent of the time of the 4-th mean New Moon in the year K.Y. exp. 3874.

Table I	3800	48.501	. . . .	16.413
Table II	74 +	1.148 +	. . . .	20.871 +
	3874	49.649		37.284
	3101 −		subtract period	29.531 −
A.D.	773		distance of first N.M. from base . .	7.753
			Table III, Sect. G, 4th N.M. . . .	88.592 +
			distance of fourth N.M. from base .	96.345
			Julian equivalent of base found . .	49.649 +
			∴ required equivalent	145.994

The year A.D. being a common year the result stands for 145 — 120 = 25 May A.D. 773, $0^{d}994$ after mean sunrise at *Laṅkā*.

NOTE: The serial numbers in brackets in Section G of Table III are only to be used in certain rare cases of true reckoning. See § 20.

§ 10. NOMENCLATURE OF LUNAR MONTHS. The lunar month which ends with the first New Moon after the *Meṣa saṃkrānti* is called *Caitra*, that which ends with the first New Moon after the *Vṛṣabha saṃkrānti* is called *Vaiśākha*, etc., as tabulated in Section A of Table III.

As however the mean synodic period of the moon — viz. 29ᵈ531 — is shorter than the distance between two mean *saṃkrāntis* — this distance being 30ᵈ438, as stated in § 6 — it happens from time to time, that a lunation which has begun shortly after a *saṃkrānti* ends before the next *saṃkrānti*. Such a *saṃkrānti*-less lunation is added to the next lunation, which obtains its regular name, according to the rule given at the beginning of this paragraph.

The two homonymous lunations are distinguished by the prefixes *prathama* (= first) and *dvitīya* (= second) or by the prefixes *adhika* (= added) and *nija* (= regular).

NOTE 1: The sidereal year evidently contains $\frac{365.258756481}{29.530587946} = 12.3688277..$ synodic periods, which implies that there must be about 369 mean added months in 1000 years. Robert Sewell, who calculated the mean intercalations for the period 3400 till 4200 of the K.Y., found within these 800 years 296 mean added months, which result is in accordance with this calculation. The fraction ·3688277.. being nearly equal to 7/19 (= 0.3684210), about the same repetitions reappear after each period of 19 years.

NOTE 2: The names of the lunar months have been derived from certain asterisms (*nakṣatras*) in the moon's track.

§ 11. MEAN ADDED MONTHS. If the distance of the first mean New Moon from the base is known, the position of all other mean New Moons with regard to all mean *saṃkrāntis* is equally known. The inferior limits determining if a month has to be added, and if so, which, are given in Section B (upper part) of Table III. We found e.g. in § 9 that the first mean New Moon of the year K.Y. exp. 3874 falls 7ᵈ753 after the base; this implies that a month *Āśvina* has to be added. By way of illustration we shall discuss another.

EXAMPLE: Is — in the mean system — a month added in the year K.Y. exp. 3926; if so, which?

We find:		☽ - ☉	
Table I	3900	19.875	
Table II	26 +	12.122 +	
	3926	31.997	
		29.531 −	
		2.466	and this being > 2.085 *Caitra* is an added month.

In fact, as the mean *Meṣa saṃkrānti* falls 32ᵈ523 after the base (and therefore the mean *saṃkrānti* preceding it 32.523 — 30.438 = 2ᵈ085 after the base) and the second New Moon 2.466 + 29.531 = 31ᵈ997 after the base, the year contains a lunation without a *saṃkrānti*, which becomes an added lunation.

NOTE: Instead of added month or lunation, the term intercalated (*prakṣipta*) is often used in chronological treatises.

§ 12. THE SERIAL NUMBER OF A LUNATION. We shall call a year with no month added a common year. A common year contains 12 lunar months, the serial numbers of which are the same of those of the *saṃkrāntis*. We find these serial numbers in the top row of Section A of table IV.

But in the case when the year contains an added month, the serial numbers of the lunations show a certain shift. E.g. when the year contains an added *Caitra*, the first lunation of that year is *adhika Caitra* (cf. the example after § 11 above), the second *nija Caitra*, the third *Vaiśākha*, etc. These serial numbers are given in Section A of Table IV. The number, given in days and decimals of a day, which has to be added to the distance of the first New Moon from the base to find the beginning of the successive months is always found in Section G of table III, headed „Multiples of synodic period of the Moon".

As an example, we shall calculate the New Moon marking the beginning of the month *Kārttika* in the expired years of the K.Y. 3873 and 3874; the first of these two years is a common year, the second contains an added *Āśvina* (See § 11):

EXAMPLE: Required the Julian date of the mean New Moons, marking the beginning of the month *Kārttika* for the years K.Y. exp. 3873 and 3874.

	base	☽ − ☉	
3800	48.501	16.413	
73 +	1.889 +	2.232 +	
3873	50.390	18.645	year common
3101 −			
772 A.D.			
Kārttika,			
8-th lunation		206.714 +	
		225.359	
base		50.390 +	
		275.749	
October		274. −	(leap year)
date		1.749	

	base	☽ − ☉	
3800	48.501	16.413	
74 +	1.148 +	20.871 +	
3874	49.649	37.284	
3101 −		29.531 −	
773 A.D.		7.753	*Āśvina* added
9-th lunation		236.245 +	
		243.998	
		49.649 +	
		293.647	
		273. −	(common year)
		20.647	

§ 13. DAYS AND *TITHIS*. A mean lunation, that is, the time elapsing between two consecutive mean New Moons, is divided into 30 *tithis*; all mean *tithis* have the same duration of $^1/_{30}$ of the synodic period, therefore of $0\overset{\mathrm{d}}{.}984$.

The days of the lunar months derive their serial numbers from those of the *tithis*, in that the day gets the serial number of the *tithi* which is current (i.e. which has already begun) at the moment of the sunrise which marks the beginning of that day.

A mean *tithi* however is $1.000 - 0.984 = 0\overset{\mathrm{d}}{.}016$ shorter than a day; if therefore a mean *tithi* begins $< 0\overset{\mathrm{d}}{.}016$ after mean sunrise, it will end before the next sunrise, and as it is not current at any sunrise cannot convey its serial number to a day. E.g. if the third *tithi* of a certain month begins shortly after sunrise and ends before the next sunrise, the days of that month will be counted: 1, 2, 4, 5 . . etc. A *tithi* which does not convey its serial number to a day of the month is called a lost (*kṣaya*) *tithi*.

The *tithis* of each month are counted in two groups; the first fifteen forming together the bright half of the month (*śukla pakṣa*), the second fifteen the dark half (*kṛṣṇa pakṣa*). The *tithis* of both halves are distinguished by their sanskrit numerals, with the exception of the fifteenth of the bright half, which ends with the Full Moon and is therefore called *pūrṇimā*, and the fifteenth of the dark half, with ends with the New Moon and is called *amāvāsyā*. The *tithi amāvāsyā* always gets 30 as its serial number (instead of *kṛṣṇa* 15).

The names of the *tithis* are to be found in columns 1 of Section B of Table IV.

NOTE: A sidereal year contains $\frac{365.2587565}{0.9843529} = 371.064$ tithis, or 5.805 more tithis than days, which implies that the number of *kṣaya tithis* in the mean system must always be 5 or 6 in each year.

§ 14. CALCULATION OF THE TIME OF BEGINNING (and ending) OF A MEAN *TITHI*. Section B of Table IV gives the numbers to be added to the distance of the mean New Moon from the base to get the times of beginning of the mean *tithis* reckoned from the base. By adding to the sum the number called the „base" of the year, we find the time the *tithi* begins according to the Julian calendar.

EXAMPLE: Required the Julian equivalend of the time of beginning of the *tithi saptamī kṛṣṇa Māgha* K.Y. exp. 3565.

	3500	45.874	6.028	
	65 +	1.819 +	0.774 +	
	3565	47.693	6.802	the year contains an added *Bhādrapada*, which
	3101 −		324.836	implies that *Māgha* is the 12-th lunation.
A.D.	464		20.671	*tithi* 7 *kṛṣṇa*.
			352.309	
			47.693 +	base.
			400.002	
Leap year, Febr.			397. −	
			3.002	A.D. 465.

The result is now that the 7-th *tithi* of the dark half of the month *Māgha* of the year K.Y. exp. 3565 begins on the third day of February of the year A.D. 465, 0^d002 after mean sunrise *Laṅkā*. As this is less than 0^d016 after sunrise, the *tithi* will end before the next sunrise, and therefore cannot convey its serial number (7) to a day of the month. The days of the month *Māgha* are now numbered: . . . . 4, 5, 6, 8, 9, 10 . . etc. of the dark half.

NOTE: Each decimal reckoning is an approximation; the last figure is always uncertain. If we had therefore found, for the beginning of the *tithi*, 0^d001 instead of 0^d002 after sunrise, our tables would have told us that either the 7-th or the 6-th of the dark half of *Māgha*, K.Y. exp. 3565 had to be considered a *kṣaya* one.

§ 15. *KARAṆAS*. In addition to the division of the lunar month into *tithis*, the *Sūrya Siddhānta* also knows of a division into *karaṇas*. A *karaṇa* is defined as the time which the moon needs to travel 6° from the sun. A mean *karaṇa* is therefore the $^1/_{60}$th part of the synodic period; the names of the *karaṇas* and the numbers to be added to the distance of a New Moon from the base to ascertain the moment at which they start are given in columns 2 and 3 of Section B of Table IV.

The Hindu calendars or *pañcāṅgas* note the ending moments of the *karaṇas*, but as a rule only of those which are current at sunrise.

EXAMPLE: Using the figures obtained in the example after § 14 we note that the *karaṇa vaṇija* was current at sunrise on the third of February A.D. 465. It ended 0ᵈ.002 after mean sunrise of that day.

LUNISOLAR RECKONING. TRUE SYSTEM

§ 16. MEAN ANOMALY AND EQUATION OF THE CENTRE

In the true system the times when the *tithis* and the *karaṇas* begin are derived from the values found in the mean system by applying two corrections, which are called: the equation of the centre of the sun, and the equation of the centre of the moon.

The equation of the centre of the sun is a function of the sun's mean anomaly, the equation of the centre of the moon is a function of the moon's mean anomaly. The values of the anomalies at the bases are found by means of columns D and E of the Tables I and II, the corresponding values of the equations are found on the folding leaves, those for the sun on the left hand, those for the moon on the right hand one.

The anomalistic period of the sun is practically equal to its sidereal period (cf. § 4 Note), viz. 365ᵈ.259; the anomalistic period of the moon is 27ᵈ.555; as soon as values for the anomalies surpassing these numbers appear in our calculation, they have to be decreased by the amounts given. To find the equations of the centre with a sufficient degree of accuracy it is necessary to work to one decimal place in the values for the sun's anomaly and to two decimal places in the values for the moon's anomaly.

The equations of the centre are positive or negative; for convenience' sake, to prevent the alternation of additions and subtractions, the negative values have been replaced by their arithmetical complements, which necessitates the subsequent subtraction of a unit; in other words: instead of subtracting x, we add $(-x + 1)$ and afterwards subtract 1 from the sun. As the absolute value of the equation never surpasses 0ᵈ.5 this cannot give rise to confusion, and it greatly facilitates the reckoning.

The Tables of the equations of the centre give values for each whole day of the sun's mean anomaly and for each tenth of a day of the moon's mean anomaly. In the calculations the mean anomalies appear with one decimal more; therefore to find the equations for the intermediate values of the anomalies an interpolation is required.

If e.g. the equation of the centre is wanted for the anomaly ☽ = 12.83, we have to proceed as follows:

For an. ☽ 12.80 the equ. according to the table = 0.908 — 1
For an. ☽ 12.90 the equ. according to the table = 0.917 — 1
therefore for the an. 12.83:

$$0.908 - 1 + 0.3 \times (0.917 - 0.908) =$$
$$0.908 - 1 + 0.3 \times 0.009 =$$
$$0.908 - 1 + 0.003 = \underline{0.911 - 1}$$

The difference between two consecutive values of the equations never surpasses ± 0.010, which implies that the interpolation is always easily effected. For convenience' sake I have added a small table of proportional parts, in which the unit stands for the third decimal. I advise careful interpolating.

EXAMPLE: Required the moment of beginning of the 10-th *tithi* of the dark half of the 10-th lunation of the year K.Y. exp. 5037.

	K.Y.exp.	base	☽ − ☉	An. ☉	An. ☽
Table I	5000	59.009	28.422	71.7	4.38
Table II	37 +	1.574 +	10.435 +	—.— +	12.82 +
	5037	60.583	38.857	71.7	17.20
	3101 —		period 29.531 —	→ 298.7 +	298.73 +
A.D.	1936		9.326	370.4	315.93
Table III, Section G 10th lunation			265.775	period 365.3 —	303.10 —
Table IV, Section B *tithi* 10 *kṛṣṇa*			23.624 +	An ☉ 5.1	An ☽ 12.83
Mean beginning of tithi			298.725		
With argument 5.1 find equ. of the centre ☉			0.016		
With argument 12.83 find equ. of the centre ☽			0.911 — 1 +		
			298.652		
Base, found above			60.583 +		
True beginning of tithi			359.235		
Table III, Section E, leap year . .			335 —		
Result: A.D. 1936, December			24	0d235 after mean sunrise mean *Laṅkā* time.	
Table III, Section F, Gregorian calendar			13 +		
Dec.			37 = January 6 A.D. 1937.		

NOTE 1: The *Sūrya Siddhānta* assumes that the sun moves in a circular orbit, the earth in its centre, at a speed which varies from moment to moment but sways round a mean value. To account for this variability of velocity and to render the

calculation of the sun's true place in its orbit possible for any moment, the *Siddhānta* accepts two points moving in the same orbit with different, but for either of them constant, speeds, in the same (easterly) direction. The first of these two points is called the *mandocca* (which we shall render by a p s i s in accordance with the editors of the translation by Burgess), the other the m e a n s u n. The apsis completes its revolution in more than 11-million years, the mean sun in a period of $365^{d}.258756481$, which period is called a sidereal year. At the end of the creation the sun, the mean sun, and the *mandocca* were in the same point of the orbit, which point is situated in the intersection of the orbit, with a straight line which joins the immovable earth with a certain point in the skies; this zero-point of the sphere is situated near the principal star of the asterism *Revatī*, which we call now *ζ-Piscium*.

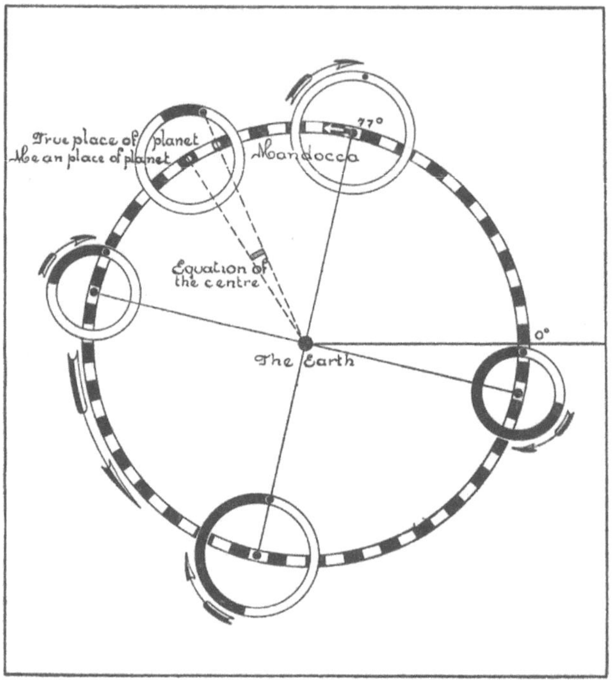

In the figure the dimensions of the epicycles and of the amount of contraction in the odd quadrants have been exaggerated.

After each sidereal year the apsis advances a fraction of a second in the orbit, and when after millions and millions of years the Kali Yuga began, the mean sun was in the zero-line and the apsis had completed a certain number of revolutions (175) plus 77° of another revolution.

The apsis is attached to the sun by cords of air and, according to its nearness, it draws the sun backward or forward; the distance of the sun from the mean never surpasses 2°10′31″. It is this deviation of the sun's place from that of the mean sun which is called e q u a t i o n o f t h e c e n t r e. To calculate this equation for any given moment, the *Sūrya Siddhānta* avails itself of an epicyclic system; in a circle having a radius of $^{14}/_{360}$ of that of the sun's orbit and having the mean sun as its centre, a point revolves at constant speed. The time of its revolution is equal to that elapsing between two consecutive passages of the mean sun through the apsis (viz. the anomalistic period) and its direction is opposite to that of the mean sun in the orbit. The point of intersection of the line joining the earth and this point revolving on the epicycle with the orbit marks the true place of the sun.

The calculation is complicated by the next assumption, viz. that the dimensions of the epicycle undergo a contraction which reaches its maximum value in the odd quadrants of the anomalistic revolution, amounting there to $^{1}/_{42}$ of the value in the even quadrants.

The position of the directional point in the epicycle is found by a simple goniometric proportion; the table of sines, however, which the *Sūrya Siddhānta* contains, differs considerably from that of the natural sines, the chief difference being that the values are only given for each 225′ in the quadrant, the others being found by linear interpolation.

The true places of the moon are determined in a similar way; the dimension of the epicycle are here $^{32}/_{360}$ with a contraction to $^{1}/_{96}$ of this amount. The anomalistic period is $27^{d}555$.

The radius of the suns' orbit is accepted to be 13.36 times that of the moon's orbit For particulars about the construction of the tables and about the formulae used in the calculation of the tables of the equations of the centre I refer to parts 1 and 2 of my article on Hindu Chronology.

NOTE 2: It follows from the text of this paragraph that the values for the equations of the centre must be found with the arguments: mean anomalies of sun and moon for the moment of true beginning of the *tithi*. But as we do not know this moment beforehand (else we should not need to calulate it) we use the moment of mean beginning. The example of the calculation given at the end of the paragraph has therefore the character of a first approximation. As a matter of fact, this first approxiamtion is amply sufficient in most cases. If, however, a greater degree of accuracy is desired, we can come a little nearer by entering the result of this approxiamtion in our calculation. E.g. we found for the total correction to be applied to the mean value, in the last example, 0.016 + 0.911 — 1 = 0.927 — 1. Applying this value to the anomalies found for the mean beginning of the *tithi*, they have to be corrected to resp.: 5.1 + 0.9 — 1 = 5.0 for the sun and 12.83 + 0.93 — 1 = 12.76 for the moon. The corresponding equations of the centre are now 0.016 (unaltered) and 0.904 — 1 (instead of 0.911 — 1); the total equation now becomes 0.920 — 1; the distance of the true beginning of the *tithi* from the base; 298.725 + 0.920 — 1 = 298.645, and the Julian equivalent 359.228. A second repetition is hardly ever of any value.

The equations of the centre have from the nature of things always to be read from the mean values.

NOTE 3: From a chronological point of view the substitution for the mean calendaric system of one based on the true movements of the sun and the moon, was anything but an improvement, as it destabilized the foundations of the timereckoning. Indeed, the system may have had the charm of adapting daily life as nearly as the astronomical knowledge permitted to the movement of the heavenly bodies, but on the other hand it broke the ties with history, as there was no unity either of elements or systems. The very complexity of the system is a proof of its primitiveness.

The transition from the mean system to the true occurred about A.D. 1000.

§ 17. *BĪJA*. The values for the moon's mean anomaly are often corrected by applying to them a correction called *bīja*, which is based on a slightly different assumption for the period of the moon's anomalistic revolution. It was not introducted before about 4500 of the *Kali Yuga*. In our Table I its amount is given as if it had existed from the beginning, to give an insight into its progress.

§ 18. DURATION OF TRUE LUNAR MONTHS. The joint effect of the two equations, that of the sun and that of the moon, causes the lunar months to be of unequal length. Calculated with the data of the *Sūrya Siddhānta* this duration is found to lie between the limits $29^{d}305$ and $29^{d}812$.

The time elapsing between two consecutive true *saṃkrāntis* varies from

$29^{d}318$ to $31^{d}644$. Accordingly, it is possible in the true system for a lunar month to remain without a *saṃkrānti*, as well as to contain two *saṃkrāntis*. In the first case a lunation is added in a similar way to that we have described already when explaining the mean system (§ § 10 and 11).

In the second case a month is suppressed.

The months *Pauṣa* and *Māgha* never appear as added months, whilst no other months can be expunged but *Mārgaśīrṣa*, *Pauṣa* and *Māgha*. *Phālguna* occasionly figures as an added month but only in years from which a month has been suppressed.

We shall treat of the true intercalations and suppressions of months in detail in the two following paragraphs.

§ 19. TRUE ADDED MONTHS. The variability in the duration of the lunar months renders it impossible to tell with certainty from the value found for ☽ – ☉ at the base of a given year if a month has to be intercalated in that year and if so, which. Only the inferior limits determining the possibility of a certain month's being intercalated can be given; these limits are tabulated in the lower part of Section B of Table III. E.g. if we find for a certain year that ☽ – ☉ at the base amounts to 6.100 it is highly probable that a month *Śrāvaṇa* has to be added to that year, it is possible that not *Śrāvaṇa* but *Āṣāḍha* has to be intercalated, but it is impossible that the year is to contain an additional *Bhādrapada*. To make sure, the exact determination of the distance of one true New Moon from the base as mentioned in Section C of Table III is wanted for each month. In the case under consideration a true New Moon occurring $124^{d}354$ after the base would show *Śrāvaṇa* to be the added month but one occurring $124^{d}350$ after the base would indicate *Āṣāḍha*.

EXAMPLE: Is a month added — and if so which — in the year K.Y. exp. 4899?

	☽ – ☉	An. ☉	An. ☽	
4800	21.499	71.7	27.42	
99	14.353	–.–	8.98	
4899	35.852	124.4	124.44	
	29.531	196.1	160.84	
	1) 6.321		137.77	1) Intercalation of *Śrāvaṇa* possible.
	2) 118.122		23.07	2) Find in Section G of Table III the number, which added to 1) brings the sum as near as possible to 124.352.
	124.443			
	0.959 — 1			
	0.353			

124.755, this being > 124.352 a lunation *Śrāvaṇa* has to be added. A second approximation (See § 16 Note 2) is not needed; it becomes necessary when the result differs less thans $0^{d}03$ from the limit.

§ 20. TRUE INTERCALATION OF *CAITRA*. A true New Moon soon after the base determines an intercalation of *Caitra*. If therefore

☽ − ⊙ at the base is found to be a little more than 0 (see the limits in the lower part of Section G of Table III), the joint effect of the two equations may cause the true New Moon to fall just after or just before the base (which we recollect to be the true *Mīna saṃkrānti*); in the first case *Caitra* is intercalated, in the second case *Phālguna* of the preceding year (which implies besides the suppression of a month, as will be shown in the next paragraph).

But it is also possible for a mean New Moon to fall just before the base; we find then ☽ − ⊙ nearing 29.531. Again the joint effect of the two equations may cause the true New Moon to occur now before or soon after the base. The first of these two cases determines an intercalation of *Phālguna* of the preceding year, the second however an intercalation of *Caitra*. To attain certainty here, we might calculate the last true New Moon of that preceding year; we gain our object sooner however by calculating the exact moment of the true New Moon, derived from a mean New Moon preceding the first mean New Moon of the year (as shown by ☽ − ⊙) by $29^{d}531$. To prevent working with negative numbers we add instead: $0^{d}469 - 30^{d}$.

If in this case a true New Moon is found soon after the base the year contains an intercalary *Caitra* and shows the peculiarity that its first mean New Moon falls before the base; we have to use in such a year those serial numbers for the synodic periods which are shown in brackets in the first column of Section G of Table III.

NOTE: The last case is a rare one; it occurs only in the years following those marked with an asterisk in Section A of the first auxiliary Table.

EXAMPLES:

Case 1. K.Y. exp. 4642	☽ − ⊙	An.⊙	An.☽
4600	14.575	71.7	22.92
42	15.038 +	–.–	20.51
		0.1 +	0.08 +
	29.613	71.8	43.51
	29.531 −		27.55 −
	0.082		15.96
	⊙ 0.169		
	☽ 0.198 +		
	0.449 *Caitra* intercalated.		

Case 2. K.Y. exp. 4379	☽ − ⊙	An.⊙	An.☽
4300	4.191	71.7	2.38
79	25.473 +	–.–	5.78
		0.1	0.13
	29.664	71.8	8.29
	29.531 −		
	0.133		
	⊙ 0.169		
	☽ 0.607 — 1 +		
	0.909 — 1	*Caitra* not intercalated (but *Phālguna* of preceding year, cf. 1st auxiliary Table, Sect. A).	

Case 3. K.Y. exp. 3628			
3600	9.490	71.7	0.38
28	19.869 +	-.-	4.49
		0.8 — 1 +	0.83 — 1 +
	29.359	71.5	4.70
☛	0.469 — 30 +		
	0.828 — 1		
	⊙ 0.169		
	☽ 0.636 — 1 +		
	0.633 — 1	*Caitra* not intercalated (but *Phālguna* of preceding year, cf. 1st auxiliary Table, Sect. A).	
Case 4. K.Y. exp. 3525			
3500	6.028	71.7	11.91
25	23.013 +	-.-	10.90
		0.5 — 1 +	0.51 — 1 +
	29.041	71.2	22.32
☛	0.469 — 30 +		
	0.510 — 1		
	⊙ 0.168		
	☽ 0.385 +		
	0.063	*Caitra* intercalated.	

§ 21. TRUE SUPPRESSIONS OF MONTHS. The values for ☽ - ⊙ at the base which serve as limits for the eventual intercalation of *Āśvina* and following months, and for the suppression of months, show only small differences, and can even overlap each other.

If we find, therefore, that ☽ - ⊙ at the base for any year lies between 10.0 and 11.50 we have to determine a series of true New Moons to establish the sequence of months in that year. This work is not difficult but it requires time. To prevent this trouble I collected in a special table (First auxiliary Table, Section A) all the years between K.Y. 3100 end 5300 (A.D. 0 till 2000) from which a month has to be expunged. This table I have good reason for believing to be correct and exhaustive.

A year from which a month has been expunged always contains one of the three months *Āśvina*, *Kārttika* or *Mārgaśīrṣa* as an added month and may contain besides an intercalary *Phālguna*. *Mārgaśīrṣa* and *Phālguna* never appear as added months in a year from which no month is expunged.

It was for these reasons that I distinguished the months *Āśvina*, *Kārttika* and *Mārgaśīrṣa* in Section B of Table III by the sign ! and put *Mārgaśīrṣa* in brackets.

EXAMPLES: I give the complete calculation for two years of different type for which ☽ - ⊙ at the base is found to lie between 10 and 11.50 to wit: 3608 and 4801:

3608	☽ - ⊙	An. ⊙	An. ☽
3600	9.490	71.7	0.38
08 +	1.458 +	-.-	1.28 +
3608	10.948	71.7	1.66

Calculate the true New Moons beginning with the one determining an intercalation of *Āśvina.*

	10.948	10.948	10.948	10.948	10.948	10.948	10.948
	177.184 +	206.714 +	236.245 +	265.775 +	295.306 +	324.836 +	354.367 +
	188.132	217.662	247.193	276.723	306.254	335.784	365.315
⊙	0.827—1	0.827—1	0.872—1	0.948—1	0.039	0.119	0.169
☽	0.269 +	0.104 +	0.919—1 +	0.751—1 +	0.632—1 +	0.586—1 +	0.624—1 +
	188.228	217.593	246.984	276.422	305.925	335.489	365.108
	Intercalation of *Āśvina* possible	but *Kārttika* is the intercalated month	*Mārgaśīrṣa* not expunged	*Pauṣa* not expunged	*Māgha kṣaya*	*Phālguna* repeated	

An. ⊙	188.1	217.7	247.2	276.7	306.3	335.8	365.3
	71.7 +	71.7 +	71.7 +	71.7 +	71.7 +	71.7 +	71.7 +
	259.8	289.4	318.9	348.4	378.0	407.5	437.0
					365.3 −	365.3 −	365.3 −
					12.7	42.2	71.7

An. ☽	188.13	217.66	247.19	276.72	306.25	335.78	365.32
	1.66 +	1.66 +	1.66 +	1.66 +	1.66 +	1.66 +	1.66 +
	189.79	219.32	248.85	278.38	307.91	337.44	366.98
	165.33 −	192.88 −	247.99 −	275.55 −	303.10 −	330.66 −	358.21 −
	24.46	26.44	0.86	2.83	4.81	6.78	8.77

4801	☽ - ⊙	An. ⊙	An. ☽
4800	21.499	71.7	27.42
01 +	18.639 +	-.-	7.05 +
4801	40.138	71.7	34.47
	29.531 −		27.55 −
	10.607		6.92

Calculate the true New Moons, again beginning with the one determining an intercalation of *Āśvina.*

	10.607	10.607	10.607	10.607	10.607	10.607
	177.184 +	206.714 +	236.245 +	265.775 +	295.306 +	324.836 +
	187.791	217.321	246.852	276.382	305.913	335.443
⊙	0.827—1	0.827—1	0.871—1	0.947—1	0.038	0.119
☽	0.831—1 +	0.684—1 +	0.599—1 +	0.594—1 +	0.670—1 +	0.810—1 +
	187.449	216.832	246.322	275.923	305.621	335.372
	Intercalation of *Āśvina* possible	Intercalation of *Kārttika* impossible; *Āśvina* is the intercalated month	*Mārgaśīrṣa* not expunged	nor *Pauṣa*	nor *Māgha* *Phālguna* not intercalated	

An. ☉	187.8 71.7 + 259.5	217.3 71.7 + 289.0	246.9 71.7 + 318.6	276.4 71.7 + 348.1	305.9 71.7 + 377.6 365.3 — 12.3	335.4 71.7 + 407.1 365.3 — 41.8
An. ☽	187.79 6.92 + 194.71 192.88 — 1.83	217.32 6.92 + 224.24 220.44 — 3.80	246.85 6.92 + 253.77 247.99 — 5.78	276.38 6.92 + 283.30 275.55 — 7.75	305.91 6.92 + 312.83 303.10 — 9.73	335.44 6.92 + 342.36 330.66 — 11.70

Inspection of Section A of the first auxiliary Table makes all calculations for the year 3608 unnecessary and reduces those for the year 4801 to the determination of the first two true New Moons.

If there are only two consecutive New Moons to be calculated the process may be shortened a little thus:

☽ - ☉	An. ☉	An. ☽	
10.607 177.184 + 187.791 29.531 + 217.322	71.7 187.8 + 259.5 29.5 + 289.0	6.92 187.79 + 194.71 192.88 — 1.83 1.98 + 3.81	being 29.531 — 27.555 cf. Section D of Table III.

	1st true N.M.	2nd true N.M.
	187.791	217.322
☉	0.827—1	0.827—1
☽	0.831—1 +	0.683—1 +
	187.449	216.832

NOTE: As perhaps the reader may wish to have the complete order of the serial numbers of the months for different types of years, I add here a schedule containing the serial numbers for a common year (cf. § 12), and for the two years which we have investigated in the two examples just given. This schedule is only an illustration of how to apply the table given in Section A of Table IV.

N°	Comm. Year	3608	4801
1	*Caitra*	*Caitra*	*Caitra*
2	*Vaiśākha*	*Vaiśākha*	*Vaiśākha*
3	*Jyeṣṭha*	*Jyeṣṭha*	*Jyeṣṭha*
4	*Āṣāḍha*	*Āṣāḍha*	*Āṣāḍha*
5	*Śrāvaṇa*	*Śrāvaṇa*	*Śrāvaṇa*
6	*Bhādrapada*	*Bhādrapada*	*Bhādrapada*
7	*Āśvina*	*Āśvina*	*Āśvina*
8	*Kārttika*	*Kārttika*	*Āśvina* II
9	*Mārgaśirṣa*	*Kārttika* II	*Kārttika*
10	*Pauṣa*	*Mārgaśirṣa*	*Mārgaśirṣa*
11	*Māgha*	*Pauṣa*	*Pauṣa*
12	*Phālguna*	*Phālguna* I	*Māgha*
13		*Phālguna* II	*Phālguna*

§ 22. TRUE *TITHIS*. A *tithi* is the time, which the moon needs to travel 12° from the sun. A true *tithi* conveys its serial number to the weekday in the manner of the mean *tithi* (§ 13), viz. the day of the month gets its serial number from that *tithi* which is current, i.e. which has already begun, at the sunrise marking the beginning of the day. Calculated from the data of the *Sūrya-Siddhānta*, the duration of the shortest *tithi* is found to be $0^{d}896$ and of the longest, $1^{d}091$.

It is therefore possible for a *tithi* beginning shortly after sunrise to end before the next sunrise; such a *tithi*, on which the sun does not rise, cannot convey its serial number to a day and *e.g.* a day 3 of a month is followed by a day 5. As we have seen when treating of the mean *tithis*, such a *tithi* is called a lost (*kṣaya*) *tithi*.

But in the true system it may also happen that a *tithi* which has begun shortly before sunrise lasts till after the following sunrise; it conveys its serial number to two consecutive days of the month and *e.g.* a day Monday No. 4 is followed by a day Tuesday No. 4. Such a *tithi* is called a repeated (*adhika*) *tithi*.

The calculation of the beginning of a true *tithi* has already been described in the example given with § 16.

It is impossible to give mean limits for the suppression or repetition of true *tithis*, that is to say: the value found for ☽ − ☉ at the base gives no clue for the distribution of the *tithis* in the course of the year. We have always to calculate the exact moment of beginning of the *tithi*, and in cases where we wish to make sure of a repetition or omission, the end as well. The end of one *tithi* is the beginning of the next. We can only state that a true *tithi*:

beginning more than $0^{d}103$ after sunrise cannot end before the next sunrise, which implies that it cannot be expunged,

beginning less than $0^{d}909$ after sunrise cannot end after the sunrise of the following day, which implies that it cannot be repeated.

EXAMPLES:

I. Required the Julian equivalent of the beginning of *tithi śukla* 13, month *Āṣāḍha*, K.Y. exp. 3585.

K.Y.exp.	base	☽ − ☉	An. ☉	An. ☽
3500	45.874	6.028	71.7	11.91
85 +	1.994 +	19.184 +	—.—	20.51
3585	47.868	25.212	125.6 +	125.62 +
3101 −	*Āṣāḍha*, 4th month	88.592	197.3	158.04
A.D. 484	*tithi* 13 *śukla*	11.812 +		137.77 −
		125.616		20.27
	☉	0.955 − 1		
	☽	0.412 +		
		125.983		
	base	47.868 +		
		173.851		
	leap year	152.		

A.D. 484, June $21{,}0^{d}851$ after mean sunrise mean *Laṅkā* time.

II. An *adhika tithi: Tithi śukla* 2 *Vaiśākha,* K.Y. exp. 5025.

	5000	59.009		28.422	71.7	4.38
	25 +	1.469 +		23.013 +	—.–	10.90
	5025	60.478		51.435	52.4 +	52.42 +
	3101 –			29.531 –	124.1	67.70
A.D.	1924			21.904		55.11 –
		Vaiśākha 2nd month		29.531		12.59
		tithi śukla 2		0.984 +		
				52.419		
			⊙	0.151		
			☽	0.888 − 1 +		
				52.458		
			base	60.478 +		

True beginning of *tithi* 112.936, the fraction being > 0.909 the *tithi* might be *adhika*. To check calculate its end as well (= beginning of next *tithi*).

	52.419	124.1	12.59
1 *tithi*	0.984 +	1.0 +	0.98 +
⊙	0.150	125.1	13.57
☽	0.980 − 1 +		
	53.533		
base	60.478 +		
True ending of *tithi*	114.011		

Therefore *tithi* 2 is current at sunrise of days 113 and 114 and day 114 also receives the serial number 2 of the month *Vaiśākha.*

The *tithi* corresponds to days May 5 and 6 A.D. 1924, Gregorian style.

III. A *kṣaya tithi. Pūrṇimā* (= 15) *Vaiśākha* K.Y. exp. 5025.

	5000	59.009		28.422	71.7	4.38
	25 +	1.469 +		23.013 +	65.2 +	10.90
	5025	60.478		51.435	136.9	65.22 +
	3101 –			29.531 –		80.50
A.D.	1924			21.904		55.11 –
		Vaiśākha 2nd month		29.531		25.39
		tithi śukla 15		13.781 +		
				65.216		
			⊙	0.127		
			☽	0.197 +		
				65.540		
			base	60.478 +		

True beginning of *tithi* 126.018, the fraction being < $0^{d}103$ the tithi might be *kṣaya*. To check, calculate its ending moment as well (= beginning of next *tithi*).

	65.216	136.9	25.39
1 *tithi*	0.984 +	1.0 +	0.98 +
	66.200	137.9	26.37
☉	0.125		
☽	0.111 +		
	66.436		
base	60.478 +		
	126.914		

The *tithi* begins after and ends before sunrise on day 126; *tithi* 14 is current at sunrise of day 126 and *tithi* 16 at that of day 127, and no day in *Vaiśākha* K.Y. exp. 5025 has 15 as its serial number.

§ 23. TRUE *KARAṆAS*. A *karaṇa* is the time which the moon needs to travel 6° from the sun. The beginning and end of a true *karaṇa* are calculated in the same manner as those of the *tithi*. The values to be added to those for the mean New Moons are given in columns 2 and 3 of Section B of Table IV.

EXAMPLE: Which *karaṇa* is current at sunrise of day 10 of the month *Bhādrapada* in the year K.Y. exp. 4995?

K.Y. exp.	base	☽-☉	An. ☉	An. ☽
4900	58.133	24.960	71.7	15.90
95 +	1.582 +	28.390 +	—.–	8.34
4995	59.715	53.350	180.3 +	180.33 +
3101 −		29.531 −	252.0	204.57
A.D. 1894		23.819		192.88 −
	Bhādrap. 6th month	147.653		11.69
	Karaṇa taitila	8.859 +		
		180.331		
	☉	0.834 — 1		
	☽	0.809 — 1		
	base	59.715 +		
True beginning of *karaṇa*		239.689		

End (necessary only in close cases):

	180.331	252.0	11.69
1 *karaṇa*	0.492 +	0.5 +	0.49 +
	180.823	252.5	12.18
☉	0.833 — 1		
☽	0.852 — 1		
base	59.715 +		
	240.223		

Therefore a *karaṇa taitila* is current at sunrise of day 240, corresponding to September 10 A.D. 1894, Gregorian style.

THE AUXILIARY TABLES

§ 24. *VĀRA* or WEEKDAY. The seven day week does not appear in Indian inscriptions before the second half of the fifth century A.D. Section B of the first auxiliary Table offers a simple means of ascertaining the weekday without reducing the result to European date.

We find *e.g.* in the example at the end of § 23 that a certain *karaṇa* begins on day 239 in the year K.Y. exp. 4995. Here the number 239 stands for day No. 239 of the Julian year of which the beginning falls in the year K.Y. exp. 4995. This day is August 27 of the Julian calendar, or September 9 of the Gregorian calendar, in the year A.D. 1894; and perpetual calendars showing the weekday for any given date of the Christian calendar are to be had in abundance. But, if we do not need the European equivalent of the date, we can ascertain the weekday straight away in the following manner:

Section B, left hand part, gives for the argument 49 . . . index 7; the right hand part gives under the index 7, with the argument 95 . . . Roman numeral VII. This result means that day No. 1 of the year K.Y. exp. 4995 is a day VII. In the lower part of Section B the septuples are tabulated, augmented by 1. The serial number of the given day, 239, happens to be among these, which means that day 239 is also a day VII, according to Section C a Saturday or *śanivāra*.

This method has the additional advantage that it is the same for common years and leap years.

NOTE: The variants for the names of the weekdays in the Index to this book are chiefly borrowed from Sewell and Dikshit's Indian Calendar, page 12.

§ 25. VARIOUS ERAS. For reasons given in § 4 we have used in our tables the era called the *Kali Yuga*. This era is however only seldom used in actual inscriptions, which implies that a given year, expressed in years of another era has to be reduced first of all to an expired year of the K.Y. For the principal eras the necessary data are to be found in Section D of the first auxiliary Table, which needs little explanation. If we read *e.g.:*

Vikrama exp. 3044 (curr. 3043) *Kārttikādi* and *Caitrādi*,

this stands for:

An expired year of the *Vikrama* era is turned into an expired year of the K.Y. by adding 3044. If — in exceptional cases — the year of the *Vikrama* era were given as a current year, we should have had to add 3043 to find the expired year of the K.Y. The years of the *Vikrama* era are considered as beginning with the month *Kārttika* or *Caitra.*

If a year does not begin with *Caitra* the correspondence is meant for that part of the year which begins with the initial month mentioned. *E.g.* a date in the month *Māgha* of the current *Kārttikādi* year 100 of the *Vikrama* era corresponds to a date in the month *Māgha* of the expired year of the

Kali Yuga (100 + 3043); but a date in a month preceding *Kārttika* corresponds to a date in K.Y. exp. 3142. For the meaning of the word *kṛṣṇa* at the end of the data for some of the eras, see the description of Section F of the first auxiliary Table in § 26.

NOTE: The name of the era, the way of counting, and the beginning of the years, is hardly ever mentioned in inscriptions, which gives rise to frequent confusions. The mention of the weekday often gives a clue to the correctness of the reduction.

§ 26. *AMĀNTA* AND *PŪRNIMĀNTA* RECKONING. We assumed in all our calculations and examples that the months began at the moment of mean or of true New Moon; this is in accordance with the common usage. But months are not infrequently assumed to commence at mean or true Full Moon, especially in the Northern countries of India.

Months commencing at New Moon are called *amānta* months, those commencing at Full Moon are called *pūrṇimānta* months.

The correspondence between *amānta* and *pūrṇimānta* months is such that the *śukla pakṣas* of homonymous months are identical. In the *pūrṇimānta* scheme the *śukla pakṣa* is the second half of the month; therefore the *kṛṣṇa pakṣa* of *Caitra* in a year counted by this scheme belongs to a year preceding the year counted by the *amānta* scheme which we use in our tables. *E.g.* a date in the *kṛṣṇa pakṣa* of *Caitra* in the year K.Y. exp. 100, counted by the *pūrṇimānta* system, belongs to the year K.Y. exp. 99 when counted in the manner of our tables.

The correspondence may be immediately read off from Section F of the first auxiliary Table.

In Section D of the same table, the eras in which the *pūrṇimānta* reckoning usually obtains are denoted by the word *kṛṣṇa*. However, many variants are used.

NOTE: Intercalations and suppressions of months are calculated throughout in the *amānta* system; the correspondence of the *pakṣas* to those of the *nija* months is retained in cases where intercalations occur. The sequence of the *kṛṣṇa* and *śukla pakṣas* is therefore interrupted in a *pūrṇimānta* month by an entire *adhika* month.

§ 27. Up to this point all our calculations and examples have been expressed in mean time for the meridian of *Laṅkā*.

Mean time is the time the sundials would show if the sun travelled along the equator at unvarying speed; for all places on the same meridian the sun would rise at the same moment. When the sun rises on the meridian of *Laṅkā* it has already risen an hour before on a meridian 15° to the East of *Laṅkā*. The people living in places on that other meridian call 0^h the moment the sun rises on their meridian. Therefore 0^h *Laṅkā* mean time is 1^h for places on a meridian 15° East of *Laṅkā* etc.

The moment of beginning of a certain *tithi* is the same everywhere, but only the people living on the same meridian give this moment the same name. *E.g.* a *tithi* beginning at 0^h on the meridian of *Laṅkā* is thought to begin at 1^h by those living on a meridian which is 15° East of that of *Laṅkā*, etc., if they are all using mean time.

The sun, however, does not travel at unvarying speed, and it does not travel along the equator.

The fact that the sun's speeds is variable causes the actual sun to be always ahead of, or behind, the mean sun; the difference, expressed in minutes of time, is called the equation of time; its amount is a function of the distance of the mean sun from the apsis (see § 16 Note 1) and does not exceed about 15 minutes of time.

The fact that the sun does not travel in the equator, but in orbits parallel to it, causes the days to be of unequal lengths. In the Northern hemisphere the sun rises later in winter than in summer, which implies that for each latitude the time of actual sunrise varies as the distance of the sun from the vernal equinox; in other words, the retardation or acceleration of sunrise is a function of the sun's tropical longitude.

The Indian *pañcāṅgas* give all *tithi*-endings in true local time, and in this lies the weakest feature of their whole chronological system. The rules the *Sūrya Siddhānta* gives for calculating the time of true sunrise are exceedingly complicated and lengthy, and inapplicable in practice. Even if these rules could be reduced to a form allowing us to determine the moment of true local sunrise within a reasonable time little would be gained, as we do not know how a *pañcāṅga*-maker in bygone days acquired his knowledge of the terrestrial longitude and latitude which were required in his calculations. We only know that his methods must have been rough and may have contained errors of many degrees.

For these reasons I adopted another method in constructing the simple tables collected in the second auxiliary Table and meant for the reduction to true local time of results in mean *Laṅkā* time. It is evident that the native methods cannot have yielded results containing very gross errors, as sunrise is a phenomenon which it is not difficult to observe. My tables here are only abbreviations of modern tables as they may be found in the works of Neugebauer and Schoch, arranged for arguments derivable from the results of the mean time calculations, or to be found on any ordinary atlas.

If now our mean time calculation gives a result which differs little from the information offered by a *pañcāṅga* we wish to check, or from the data mentioned in a given inscription, *e.g.* if the inscription mentions a 4-th *tithi* as *adhika*, whilst we have found the third or the fifth, or if our answer is one day out, giving for example a Sunday where the inscription gives Saturday or Monday, we can see from this second auxiliary Table whether the discrepance may be caused by the difference between mean time *Laṅkā* and true local time. If this proves to be the case, we are justified in accepting the information of the *pañcāṅga* or the inscription as correct. This is all we can do; Hindu chronology is not free from a certain amount of uncertainty. This does not apply to the intercalations and omissions of months; if the *Siddhānta* that has been followed is known, these can be established without a shadow of doubt. As sunrise does not enter in the calculations of intercalations and expunctions, they must be the same everywhere in the world.

To turn the time when a *tithi* begins, determined by our tables in mean *Laṅkā* time, into true local time, we use Sections A—D of the second auxiliary Table.

EXAMPLE: We found that a true *tithi* began in K.Y. exp. 3585 on day 173.851 (cf. example 1 in § 22) expressed in mean time *Laṅkā*. What is the beginning of that same tithi in true local time for Eran, when the longitude of that place is 78°40' East of Greenwich, and its latitude 24°?

We find in the second auxiliary Table:

in Section A at the argument 78 2/3	+ 0.008
in Section B at the arguments 174 and 3600	+ 0.000
in Section D at the arguments (174 — 5) and 24°	+ 0.034
The number △ = — 5 has been found in Section C with the argument 3600	
Total equation	0.042
Mean beginning	173.851
Beginning of *tithi* in true local time at Eran	173.893

EXAMPLE 2: A *tithi* ended in K.Y. 5011 on day 182.876; when does it end at Madras (lat. 13°, long. 80° E. of Gr.)?

Sect. A, arg. 80	+ 0.012
Sect. B, arg. 183/5000	— 0.004
Sect. D, arg. 13/(183 + 5); △ = 5 acc. to Sect. C	+ 0.017
Total equation	0.025
Mean end of *tithi*	182.876
End of *tithi* in true local time at Madras	182.901

NOTE: The above examples have been chosen for comparison, as they appear in modern works on Hindu chronology. Venkatesh and Swamikannu both find for the total equation in Ex. 1 $0^{\mathrm{d}}039$, although they do not quite agree as to the coordinates of Eran. In the second example Swamikannu finds $0^{\mathrm{d}}025$, whilst his final result differs again $0^{\mathrm{d}}015$ from the information the Madras „College Panchang" gives for that year.
Apart from special cases I advise the reader not to aim at closer figures for the determination of true local sunrise than our second auxiliary Table gives.

PRACTICAL EXERCISES

The answers are on page 33.

1. (§ 5). Find the base for K.Y. exp. 3029.
2. What does the answer to the first question stand for?
3. (§ 6). Find the true Kumbha *saṃkrānti* for K.Y. exp. 4635.
4. Find the equivalent Julian date and the time of day. (see Aux. Table II, Sect. E).
4. Find the equivalent Julian date and the time of day.
5. Find the Gregorian equivalent and the time (in *ghaṭikās* and *palas* [see aux. Table II, sect. E]) of the mean *Mīna saṃkrānti* in K.Y. exp. 4932.
6. (§ 7). Find the Julian equivalent of 24 *Karka* K.Y. exp. 4372, using the true *saṃkrānti* and the *Orissa* rule.
7. (§ 9). Find the distance of the first mean New Moon from the base in K.Y. exp. 5772.
8. The same for K.Y. exp. 4227.
9. Find the distance of the 11-th mean New Moon from the base in K.Y. exp. 5000.
10. Find the Gregorian equivalent of the same.
11. (§ 11). Is a mean month added in K.Y. exp. 3687; if so which?
12. Find how much time elapsed between the beginning of the mean intercalated month found above and the *saṃkrānti* immediately preceding it, and how much time elapsed between the end of the same lunation and the next *saṃkrānti*.
13. (§ 12). Find the mean New Moon marking the beginning of mean *Māgha* in K.Y. exp. 3687.
14. (§ 14). Find the Julian equivalent of the beginning of the mean *tithi* 5 *śukla Kārttika* K.Y. exp. 4035.
15. (§ 16). Find the mean anomaly of the sun for a moment 100^d0 after the first mean N.M. after the base in K.Y. exp. 1234.
16. The same for the mean anomaly of the moon in K.Y. exp. 4321.
17. Find the equation of the centre for the sun for the mean anomaly 200.0.
18. The same for the mean anomaly 200.4.
19. Find the equation of the centre of the moon for the mean anomaly 14.10.
20. The same for the mean anomaly 14.13.
21. (§ 17). Find the mean anomaly of the moon as in problem 16, this time taking the *bīja* into account.
22. (§ 19). Is it possible for a true month to be added in K.Y. exp. 5013; if so, which? Is it in fact added?
23. The same for K.Y. exp. 5008.
24. (§ 21). Is a month expunged in K.Y. exp. 4454?
25. Is a true month added in K.Y. exp. 4454? If so, which?
26. (§ 22). Find the beginning and end, and the Julian equivalents, of the true *tithi* 9 *kṛṣṇa Phālguna* K.Y. exp. 4303. To which day or days does it correspond?
27. (first aux. Table, Section B). Find the weekday corresponding to day 433 of the the Julian year commencing in K.Y. exp. 4303.
28. (*ibid.* Sect. D). Find the year K.Y. exp. corresponding to *Śaka* 1000 curr.

INDEX AND GLOSSARY

The *arabic* numerals refer to the paragraphs of the Explanation, the *roman* numerals to the Tables and Sections.
☉ = Sunday ☽ = Monday ♂ = Tuesday ☿ = Wednesday
♃ = Thursday ♀ = Friday and ♄ = Saturday.

Abjavāra	☽
added months	10
— — mean	11
— — true	19
adhiḵa	added
Adi (tamil)	*Karḵa*
Ādivāra	☉
Ādityavāra	☉
Aghran (bengali)	*Mārgaśīrṣa*
Aharpativāra	☉
Ahasḵaravāra	☉
amānta- and *pūrṇimānta* schemes — correspondence of	1st aux. Table F
— reckoning or -scheme	26
amāvāsyā	13
—	*tithi* 30 IV B
Aṅgāraḵavāra	♂
Āṅgirasavāra	♃
Ani (tamil)	*Mithuna*
anomalistic period ☽	16
— year	4 note, 16
anomaly cf. mean anomaly	
apsis	16 note 1
Arḵavāra	☉
Aruṇavāra	☉
Āṣāḍha, 4th month	III A, IV A
Aṣṭamī	*tithi* 8, IV B
Āśvina, 7th month	III A, IV A
Ati (tamil)	*Āṣāḍha*
Avani (tamil)	*Siniha*
Avantī	4 note
badi	*ḵṛṣṇa*
bahula	*ḵṛṣṇa*
Bandhavāra	☿
base	5
Bava, ḵaraṇa śuḵla 2. 9. 16. 23. 30 *ḵṛṣṇa* 7. 14. 21	IV B
Besa (tamil)	*Vaiśāḵha*
Bhadra, ḵaraṇa śuḵla 8	IV B

Bhādrapada, 6th month	III A, IV A
Bhānuvāra	☉
Bhārgavavāra	♀
Bhāskaravāra	☉
Bhaṭṭārakavāra	☉
Bhaumavāra	♂
Bengal San, era	1st aux. Table D
— rule	1st aux. Table E
bīja	17
Bontelu (*tamil*)	*Āśvina*
Bradhnavāra	☉
Bṛihaspativāra	♃
bright half	13
Bhṛiguvāra	♀
Budhavāra	☿
Caitra, 1-st month	8, III A, IV A
— , particulars of true intercalation of	20
Caitrādi	beginning with *Caitra*
Caturdaśī	*tithi* 14, IV B
Caturthī	*tithi* 4, IV B
catuṣpāda, *karaṇa kṛṣṇa* 30	IV, B
Chandramasvāra	☽
Chandravāra	☽
Chedi, era	1-st aux. Table D
common year	12
current years	3
Daityaguruvāra	♀
Dakṣiṇāyana saṃkrānti	*Karka*
dark half	13
Daśamī	*tithi* 10, IV B
day	13
Dhanus, *saṃkrānti* 9	III A
Dhiṣaṇavāra	♃
distance of mean New Moon from base	9
duration of true lunar months	18
Dvādaśī	*tithi* 12, IV B
dvitīya	second, *nija*, regular
—	*tithi* 2, IV B
Ekadaśī	*tithi* II, IV B
epicycle	16, note 1
epoch	2
— of the *Kali Yuga*	4
equation of the centre	16
— — time	27
expired years	3

expunction of months 21
— — *tithis*, mean 13
— — — true 22
expunged months — Table of . . . . . . 1st aux. Table, A
Gara, *ḳaraṇa śuḳla* 5. 12. 19. 26, *ḳṛṣṇa* 4, 11, 18, 15 IV B
gata . . . . expired
Gregorian calendar . . . . III, F
Gupta, era . . . . 1st aux. Table D
Guruvāra . . . . ♃
halfs - bright and dark 13
Induvāra . . . . ☽
intercalated . . . . added
Iṣa . . . . *Āśvina*
Jarde (*tamil*) . . . . *Kārttiḳa*
Jyeṣṭha 3d month . . . . III A, IV A
Julian calendar 24
Kali Yuga, era . . . . 3, 4 1st aux. Table D
Kanyā, *saṃḳrānti* 6 . . . . III, A
Kanyādi . . . . beginning with *Kanyā*
ḳaraṇas – mean 15
— – true 23
Karḳa, *saṃḳrānti* 4 . . . . III A
Kartelu (*tamil*) . . . . *Jyestha*
Kārttiḳa, 8th month . . . . III A, IV A
ḳaulava, *ḳaraṇa śuḳla* 4, 11, 18, 25, *ḳṛṣṇa* 2, 9, 16, 23 IV B
Kavivāra, *Kavyavāra* . . . . ♀
Kiṃstughna, *ḳaraṇa śuḳla* 1 . . . . IV B
Kollam, era . . . . 1st aux. Table D
ḳṛṣṇa pakṣa . . . . dark half
Kṣapāḳaravāra . . . . ☽
ḳṣaya . . . . lost
— months 21
— *tithis* 13, 22
Kumbha, *saṃḳrānti* 11 . . . . III A
Laṅḳā . . . . 4 note
lunation . . . . lunar month
lunisolar reckoning 8
— — , mean . . . . 9—15
— — , true . . . . 16—13
— months . . . . III A
Mādhava . . . . *Vaiśāḳha*
Madhu . . . . *Caitra*
madhyama . . . . mean
Māgha, 11th lunar month . . . . III A, IV A
Mahīsutavāra . . . . ♂

Makara, saṃkrānti 10	III A
Mandavāra	♄
mandocca	apsis
Mangalavāra	♂
Mārgaśīrṣa, 9th month	III A, IV A
Mayi (*tamil*)	*Māgha*
mean added months	11
— anomalies	I and II, D and E
— *karaṇas*	15
— reckoning	8
— *saṃkrāntis*	6
— sun and moon	note 1; 27
— sunrise	27
— time	27
— *tithis*	14
Meṣa, 1st *saṃkrānti*	III A
Meṣādi	beginning with *Meṣa*
Mīna, saṃkrānti 12	III A
— — , true = base	5
Mithuna, saṃkrānti 3	III A
months – expunction of	21
— , mean added	11
— , nomenclature of lunar	10
— , solar	7
— , true added	19
multiples of anomalistic period ☽	III H
— — synodic period ☽	III G
Nabhas	*Śrāvaṇa*
Nabhaya	*Bhādrapada*
Nāga, karaṇa kṛṣṇa 29	IV B
nakṣatras	10 note
Navanū, tithi 9	IV B
nija	regular
Nirnala (*tamil*)	*Bhādrapada*
Niṣpativāra	☽
nomenclature of lunar months	10
Orissa rule	1st aux. Table E
Paggu (*tamil*)	*Caitra*
pakṣa	13
Pañcamī, tithi 3	IV B
Pauṣa, 10th month	III A, IV A
Perarde (*tamil*)	*Mārgaśīrṣa*
Phālguna, 12th month	III A, IV A
prathama	first, *adhika*, added
Pratipadā	*tithi* 1 *śukla*, IV B

Puntelu (*tamil*)	*Pauṣa*
pūrṇimā	13
— , *tithi śukla* 15	IV B
pūrṇimānta reckoning or-scheme	26
rāśi	6
—	*kṛṣṇa*
Rauhiṇeyavāra	☿
Ravivāra	☉
repeated	intercalated, added
— true *tithis*	22
Revatī	16 note 1
Rohitāṅgavāra	♂
Sahas	*Mārgaśīrṣa*
Sahasya	*Pauṣa*
Śaka, era	1st aux. Table D
saṃkrāntis	6, III A
Śakuni, *karaṇa kṛṣṇa* 28	IV B
San, era — *Bengal*	1st aux. Table D
Śanivāra	♄
Saptamī, *tithi* 7	IV B
Saptarṣi, era	1st aux. Table D
Ṣaṣṭi, *tithi* 6	IV B
Saumyavāra	☿
Sauramāsa	solar month
Saurivāra	♄
serial numbers of lunations	IV A
sidereal year	4 note
sign of the equation of the centre	16
Siṃha, *saṃkrānti* 5	III A
Siṃhādi	beginning with *Siṃha*
solar months	7
— reckoning	6—7
Somavāra	☽
Sona (*tamil*)	*Śrāvaṇa*
spaṣṭa	true
Śrāvaṇa, 5th month	III A, IV A
Śuci	*Āṣāḍha*
śuddha, *śudi*	*śukla*
Suggi (*tamil*)	*Phālguna*
śukla pakṣa	bright half
Śukra	*Jyeṣṭha*
Śukravāra	♀
Suracharyavāra	♃
Sūrya Siddhānta	page 1, year 1860
synodic period	8

synodic period, multiples of . . . . III G
Taitila, karaṇa śukla 5, 12, 19, 26 *kṛṣṇa* 3, 10, 17, 24 IV B
Tapas . . . . *Māgha*
Tapasya . . . . *Phālguna*
time – graphical representation of 1
tithis – lost, *kṣaya* 13, 22
— mean 13, 14
— names of . . . . IV B
— true 22
Trayodaśī, tithi 13 . . . . IV B
Tṛtīya, tithi 3 . . . . IV B
tropical longitude of the sun 27
— year . . . . 4 note
true expunction of months 21
— added months 19
— intercalation of *Caitra* 20
— *karaṇas* 23
— local time 27
— reckoning 8
— repeated *tithis* 22
— *saṃkrāntis* 6
— sunrise 27
— *tithis* 22
Tulā, saṃkrānti 7 . . . . III A
Ujjayinī . . . . 4 note
Ūrja . . . . *Kārttika*
Uśanasvāra . . . . ♀
uttarāyaṇa saṃkrānti . . . . *Makara*
Vācaspativāra . . . . ♃
vadi, vadya . . . . *kṛṣṇa*
Vaiśākha, 2nd month . . . . III A, IV A
Vālava, karaṇa śukla 3, 10, 17, 24, *kṛṣṇa* 1, 8, 15, 22 IV B
Vaṇija, karaṇa śukla 7, 14, 21, 28, *kṛṣṇa* 5, 12, 19, 26 IV B
vāra . . . . weekday
vartamāna . . . . current
Vikrama, era . . . . 1st aux. Table D
Vilayati, era . . . . 1st aux. Table D
Viṣṭi, karaṇa śukla 8, 15, 22, 29 *kṛṣṇa* 6, 13, 20, 27 IV B
Vṛścika, saṃkrānti 8 . . . . III A
Vṛṣabha, saṃkrānti 2 . . . . III A
weekday . . . . 24, 1st aux. Table B
year – anomalistic . . . . 4 note
— – civil . . . . 4 note
— – common. 12
— – current 3

year – expired 3
— – sidereal . . . . 4 note
— – tropical . . . . 4 note

THE PROBLEMS ANSWERED

1. 43.000; 2. Mean sunrise in *Laṅkā* mean time of day 43 of the Julian year 3029 — 3101 = — 72; 3. 392.000; 4. January 27 A.D. 1535 at mean sunrise mean *Laṅkā* time; 5. March 13 A.D. 1832 45 gh. 25 p. after mean sunrise mean *Laṅkā* time; 6. July 21 A.D. 1271; 7. $6^{d}.715$; 8. $1^{d}.959$; 9. $323^{d}.728$; 10. January 30 A.D. 1900, $0^{d}.737$ after mean sunrise, mean *Laṅkā* time; 11. Yes; *Bhādrapada*; 12. $0^{d}.267$ and $0^{d}.641$; 13. January 14 A.D. 587, $0^{d}.988$ after mean sunrise, mean *Laṅkā* time; 14. October 14 A.D. 934, $0^{d}.983$ after mean sunrise, M.L.T.; 15. 200.4; 16. 14.13; 17. 0.947 — 1; 18. 0.946 — 1; 19. 0.031; 20. 0.034; 21. 14.24; 22. Yes. *Āṣāḍha*. Yes; 23. Yes. *Caitra*. Yes; 24. No, as shown by Section A of the first auxiliary Table; 25. Yes. *Bhādrapada* (not *Āśvina*); 26. 432.072 and 433.134 A.D. 1203, March 9; 27. Sunday; 28. K.Y. exp. 4178.

ERRATA

In the diagram opposite page 3 read in no. 21 *Kumbha* in stead of *Kumba*.

Printed by A.A.M. Stols
Maastricht (Holland)
January 1938

TABLES

A A

A

K.Y.exp.	adhika	kṣaya	adhika	K.Y.exp.	adhika	kṣaya	adhika
·3101	*āśvina*	*pauṣa*		4359	*kārttika*	*pauṣa*	
3223	*ā*	*mārgaśīrṣa*		4378	*k*	*pauṣa*	*phālguna*
·3242	*ā*	*mārgaśīrṣa*		4397	*mārgaśīrṣa*	*pauṣa*	*p*
·3364	*kārttika*	*mārgaśīrṣa*		4416	*k*	*mārgaśīrṣa*	*p*
·3383	*ā*	*mārgaśīrṣa*		4435	*āśvina*	*pauṣa*	*p*
3505	*k*	*mārgaśīrṣa*	*phālguna*	4481	*k*	*mārgaśīrṣa*	
·3524	*ā*	*pauṣa*		4500	*k*	*pauṣa*	
3589	*k*	*pauṣa*		4576	*ā*	*māgha*	*p*
3608	*k*	*māgha*	*p*	4622	*k*	*mārgaśīrṣa*	
3627	*k*	*pauṣa*	*p*	4641	*ā*	*pauṣa*	
3646	*k*	*pauṣa*	*p*	·4782	*ā*	*pauṣa*	
3665	*ā*	*pauṣa*	*p*	·4923	*ā*	*pauṣa*	
3711	*k*	*mārgaśīrṣa*		·5064	*ā*	*pauṣa*	
3730	*ā*	*pauṣa*		5083	*ā*	*māgha*	*p*
3852	*k*	*mārgaśīrṣa*		5129	*k*	*mārgaśīrṣa*	
3871	*ā*	*pauṣa*		5148	*mārgaśīrṣa*	*pauṣa*	*p*
3993	*k*	*mārgaśīrṣa*		5186	*k*	*mārgaśīrṣa*	*p*
·4012	*ā*	*pauṣa*		5205	*ā*	*pauṣa*	*p*
·4153	*ā*	*pauṣa*		5224	*ā*	*pauṣa*	*p*
·4294	*ā*	*pauṣa*		5251	*k*	*mārgaśīrṣa*	

A Table of expired years of the *Kali-Yuga* in which a month has been suppressed.

B

anni cent. *K.Y.*				anni expansi	1 2	3 4 5	6 7	anni expansi
32	39	46	3	0 11 **17** 22 28 39 **45**	I II	III IV V	VI VII	50 56 67 **73** 78 84 95
33	40	47	2	**1** 6 12 23 **29** 34 40	II III	IV V VI	VII I	51 **57** 62 68 79 **85** 90 96
34	41	48	1	7 **13** 18 24 35 **41** 46	III IV	V VI VII	I II	52 63 **69** 74 80 91 **97**
35	42	49	7	2 8 19 **25** 30 36 47	IV V	VI VII I	II III	**53** 58 64 75 **81** 86 92
36	43	50	6	3 **9** 14 20 31 **37** 42 48	V VI	VII I II	III IV	59 **65** 70 76 87 **93** 98
37	44	51	5	4 15 **21** 26 32 43 **49**	VI VII	I II III	IV V	54 60 71 **77** 82 88 99
38	45	52	4	**5** 10 16 27 **33** 38 44	VII I	II III IV	V VI	55 **61** 66 72 83 **89** 94

1	29	57	85	113	141	169	197	225	253	281	309	337	365	393	421	449
8	36	64	92	120	148	176	204	232	260	288	316	344	372	400	428	456
15	43	71	99	127	155	183	211	239	267	295	323	351	379	407	435	463
22	50	78	106	134	162	190	218	246	274	302	330	358	386	414	442	470

B

C

The vāra or weekday

Cf. the Index for a number of Sanskrit synonyms

I	☉	*Āditya-Ravivāra*
II	☽	*Somavāra*
III	♂	*Maṅgalavāra*
IV	☿	*Budhavāra*
V	♃	*Guruvāra*
VI	♀	*Śukravāra*
VII	♄	*Śanivāra*

D

Numbers to be added to the years of various *eras*, to obtain exp. years of the *Kali-Yuga:*

Kali Yuga, (current — 1); *Saptarṣi* (exp. 3026), curr. 3025 *Caitrādi; Vikrama* exp. 3044 (current 3043) *Kārttikādi* and *Caitrādi; Śaka*, exp. 3179 (curr. 3178) *Caitrādi; Gupta* (exp. 3421), curr. 3420, *Caitrādi kṛṣṇa; Chedi* (exp. 3549) curr. 3548, *Aśvinidi kṛṣṇa; Vilāyati*, (exp. 3694), curr. 3693, *Kanya-Saṃkrānti*, solar; *Bengal San* (exp. 3695), curr. 3694, *Vaiśakhādi*, solar; *Kollam* (exp. 3927), curr. 3926, *Kanyā* or *Siṃha Saṃkrānti*, solar.

E

The solar year: Different rules exist as to the beginning of the months of the solar year. If we represent the moment of a *saṃkrānti* by D + d (Day and decimals of the day), the first day of the next following month begins at sunrise of day: D + 1 (d < 0.75) or D + 2 (d > 0.75) according to the *Bengal rule* and of day D (d < 0.75) or D + 1 (d > 0.75) according to the *Orissa rule.*

E

F

Correspondance of *amānta* and *pūrṇimānta* months:

amānta reckoning: | *Caitra* | *Vaiś.* | *Jyeṣth.* | *Āśadh.* | *Śrāv.* | *Bhādr.* | *Āśvina* | *Kārtt.* | *Mārg.* | *Pauṣa* | *Māgha* | *Phālg.*

ś=śukla, k=kṛṣṇa pakṣa k ś k ś k ś k ś k ś k ś k ś k ś k ś k ś k ś k ś k

pūrnimānta reckoning: | *Caitra* | *Vaiś.* | *Jyeṣth* | *Āśadh.* | *Śrāv.* | *Bhādr.* | *Āśvina* | *Kārtt.* | *Mārg.* | *Pauṣa* | *Māgha* | *Phālg.* | *C.*

F

FIRST AUXILIARY TABLE

A	B	C	D	E	F
Expired years of the *Kali-Yuga* *Śaka exp.*+3179	Base being the Julian date of true *Mīna-Saṃkr.*	☽—☉ Distance of first mean New Moon from base	An. ☉ Mean anomaly of the Sun	An. ☽ Mean anomaly of the Moon without *bīja*	 *bīja* (additive)
0000	15.227	2.993	71.8	1.92	0.00
1000	23.983	8.079	71.8	24.45	0.03
1100	24.859	11.540	71.8	12.93	03
1200	25.734	15.002	71.8	1.41	03
1300	26.610	18.464	71.8	17.44	03
1400	27.486	21.925	71.8	5.91	0.04
1500	28.361	25.387	71.8	21.95	04
1600	29.237	28.849	71.8	10.42	04
1700	30.113	2.780	71.8	26.45	04
1800	30.988	6.241	71.8	14.93	0.05
1900	31.864	9.703	71.8	3.40	05
2000	32.739	13.165	71.8	19.43	05
2100	33.615	16.626	71.8	7.91	05
2200	34.491	20.088	71.8	23.94	0.06
2300	35.366	23.549	71.8	12.42	06
2400	36.242	27.011	71.8	0.89	06
2500	37.118	0.942	71.8	16.92	06
2600	37.993	4.404	71.8	5.40	0.07
2700	38.869	7.865	71.8	21.43	07
2800	39.745	11.327	71.8	9.91	07
2900	40.620	14.789	71.8	25.94	07
3000	41.496	18.250	71.8	14.42	0.08
3100	42.372	21.712	71.7	2.89	08
3200	43.247	25.174	71.7	18.92	08
3300	44.123	28.635	71.7	7.40	08
3400	44.999	2.566	71.7	23.43	0.09
3500	45.874	6.028	71.7	11.91	09
3600	46.750	9.490	71.7	0.38	09
3700	47.626	12.951	71.7	16.41	09
3800	48.501	16.413	71.7	4.89	0.10
3900	49.377	19.875	71.7	20.92	10
4000	50.252	23.336	71.7	9.40	10
4100	51.128	26.798	71.7	25.43	10
4200	52.004	0.729	71.7	13.90	0.11
4300	52.879	4.191	71.7	2.38	11
4400	53.755	7.652	71.7	18.41	11
4500	54.631	11.114	71.7	6.89	11
4600	55.506	14.575	71.7	22.92	0.12
4700	56.382	18.037	71.7	11.39	12
4800	57.258	21.499	71.7	27.42	12
4900	58.133	24.960	71.7	15.90	12
5000	59.009	28.422	71.7	4.38	0.13
5100	59.885	2.353	71.7	20.41	13
5200	60.760	5.815	71.7	8.88	13
5300	61.636	9.276	71.7	24.91	0.14
5400	62.512	12.738	71.7	13.39	14
5500	63.387	16.200	71.7	1.87	14
5600	64.263	19.661	71.7	17.90	14
5700	65.138	23.123	71.7	6.37	0.15
5800	66.014	26.585	71.7	22.40	15

TABLE I

A Table of the Equation of the Centre of the Sun to be used in calculating *tithis* and intercalations etc. of months.

	0.		0.		0.		0.		0.—1		0.—1		0.—1
0	000	53	142	106	173	159	072	211	915	264	824	317	868
1	003	54	144	107	172	160	069	212	913	265	824	318	870
2	006	55	145	108	171	161	066	213	910	266	823	319	872
3	009	56	147	109	170	162	063	214	907	267	823	320	874
4	013	57	149	110	170	163	060	215	905	268	823	321	876
5	016	58	150	111	169	164	057	216	902	269	822	322	878
6	019	59	152	112	167	165	054	217	899	270	822	323	880
7	022	60	154	113	166	166	051	218	897	271	822	324	883
8	025	61	155	114	165	167	048	219	895	272	822	325	885
9	028	62	157	115	164	168	045	220	892	273	822	326	887
10	031	63	158	116	163	169	042	221	890	274	822	327	890
11	034	64	159	117	162	170	039	222	887	275	822	328	892
12	037	65	161	118	160	171	036	223	885	276	822	329	895
13	040	66	162	119	159	172	033	224	882	277	822	330	897
14	043	67	163	120	158	173	030	225	880	278	822	331	900
15	046	68	165	121	156	174	027	226	878	279	822	332	902
16	049	69	166	122	155	175	024	227	876	280	823	333	905
17	052	70	167	123	153	176	021	228	874	281	823	334	908
18	055	71	168	124	151	177	018	229	871	282	823	335	910
19	058	72	169	125	150	178	015	230	869	283	824	336	913
20	061	73	170	126	148	179	011	231	867	284	824	337	916
21	064	74	171	127	146	180	008	232	865	285	825	338	918
22	067	75	172	128	145	181	005	233	863	286	825	339	921
23	070	76	172	129	143	182	002	234	861	287	826	340	924
24	073	77	173	130	141		**0.—1**	235	859	288	827	341	927
25	075	78	174	131	139	183	999	236	858	289	827	342	930
26	078	79	174	132	137	184	996	237	856	290	828	343	932
27	081	80	175	133	135	185	993	238	854	291	829	344	935
28	084	81	176	134	133	186	989	239	852	292	830	345	938
29	086	82	176	135	131	187	986	240	851	293	831	346	941
30	089	83	177	136	129	188	983	241	849	294	832	347	944
31	092	84	177	137	127	189	980	242	847	295	833	348	947
32	094	85	177	138	125	190	977	243	846	296	834	349	950
33	097	86	178	139	123	191	974	244	844	297	835	350	953
34	100	87	178	140	120	192	971	245	843	298	836	351	956
35	102	88	178	141	118	193	968	246	841	299	838	352	959
36	105	89	178	142	116	194	965	247	840	300	839	353	962
37	107	90	178	143	113	195	962	248	839	301	840	354	965
38	110	91	178	144	111	196	959	249	837	302	842	355	968
39	112	92	178	145	109	197	956	250	836	303	843	356	971
40	114	93	178	146	106	198	953	251	835	304	844	357	974
41	117	94	178	147	104	199	950	252	834	305	846	358	977
42	119	95	178	148	101	200	947	253	833	306	848	359	980
43	121	96	178	149	099	201	944	254	832	307	849	360	984
44	123	97	178	150	096	202	941	255	831	308	851	361	987
45	126	98	177	151	093	203	938	256	830	309	852	362	990
46	128	99	177	152	091	204	935	257	829	310	854	363	993
47	130	100	176	153	088	205	932	258	828	311	856	364	996
48	132	101	176	154	085	206	929	259	827	312	858	365	999
49	134	102	175	155	083	207	926	260	827	313	860		**0.**
50	136	103	175	156	080	208	924	261	826	314	862	366	002
51	138	104	174	157	077	209	921	262	825	315	863	367	005
52	140	105	174	158	074	210	918	263	825	316	865	368	009

A	B	C	D	E
Years	Base	29.531 ☽ − ☉	An. ☉	Anomaly ☽
00	1.000	0.000		0.00
01	1.259	18.639		7.05
02	0.518	7.747		14.10
03	0.776	26.386		21.15
04	1.035	15.494		0.64
05	1.294	4.603		7.69
06	0.553	23.242		14.74
07	0.811	12.350		21.79
08	1.070	1.458		1.28
09	1.329	20.097		8.33
10	0.588	9.205		15.38
11	0.846	27.844		22.43
12	1.105	16.953		1.92
13	1.364	6.061		8.97
14	0.623	24.670		16.02
15	0.881	13.808		23.07
16	1.140	2.916		2.56
17	1.399	21.555		9.61
18	0.658	10.663		16.66
19	0.916	29.302		23.71
20	1.175	18.411		3.21
21	1.434	7.519		10.26
22	0.693	26.158		17.30
23	0.951	15.266		24.35
24	1.210	4.374		3.85
25	1.469	23.013		10.90
26	0.728	12.122		17.95
27	0.986	1.230		24.99
28	1.245	19.869		4.49
29	1.504	8.977		11.54
30	0.763	27.616		18.59
31	1.021	16.724		25.64
32	1.280	5.833		5.13
33	1.539	24.472		12.18
34	0.798	13.580		19.23
35	1.056	2.688		26.28
36	1.315	21.327		5.77
37	1.574	10.435		12.82
38	0.833	29.074		19.87
39	1.092	18.182		26.92
40	1.350	7.291		6.41
41	1.609	25.930		13.46
42	0.868	15.038		20.51
43	1.127	4.146		0.00
44	1.385	22.785		7.05
45	1.644	11.893		14.10
46	0.903	1.002		21.15
47	1.162	19.641		0.65
48	1.420	8.749		7.69
49	1.679	27.388		14.74

A	B	C	D	E
Years	Base	29.531 ☽ − ☉	An. ☉	Anomaly ☽
50	0.938	16.496		21.79
51	1.197	5.604		1.29
52	1.455	24.243		8.34
53	1.714	13.352		15.38
54	0.972	2.460		22.43
55	1.232	21.099		1.93
56	1.490	10.207		8.98
57	1.749	28.846		16.03
58	1.008	17.954		23.08
59	1.267	7.063		2.57
60	1.525	25.701		9.62
61	1.784	14.810		16.67
62	1.043	3.918		23.72
63	1.302	22.557		3.21
64	1.560	11.665		10.26
65	1.819	0.774		17.31
66	1.078	19.412		24.36
67	1.337	8.521		3.85
68	1.595	27.160		10.90
69	1.854	16.268		17.95
70	1.113	5.376		25.00
71	1.372	24.015		4.49
72	1.630	13.123		11.54
73	1.889	2.232		18.59
74	1.148	20.871		25.64
75	1.407	9.979		5.13
76	1.665	28.618		12.18
77	1.924	17.726		19.23
78	1.183	6.834		26.28
79	1.442	25.473		5.78
80	1.701	14.582		12.82
81	1.959	3.690		19.87
82	1.218	22.329		26.92
83	1.477	11.437		6.42
84	1.736	0.545		13.47
85	1.994	19.184		20.51
86	1.253	8.292		0.01
87	1.512	26.931		7.06
88	1.771	16.040		14.11
89	2.029	5.148		21.16
90	1.288	23.787		0.65
91	1.547	12.895		7.70
92	1.806	2.003		14.75
93	2.064	20.642		21.80
94	1.323	9.751		1.29
95	1.582	28.390		8.34
96	1.841	17.498		15.39
97	2.099	6.606		22.44
98	1.358	25.245		1.93
99	1.617	14.353		8.98

TABLE II

	A			B		C	
	The *Saṃkrāntis* true and mean with the lunisolar months *ending* after them			Inferior limits for the intercalation of months ☽ − ☉ at base:		Check for true intercalations and suppressions of lunisolar months	
	Meṣa	T	30.354	Mean	certain	! Suppressions of months are possible when ☽ − ☉ at base is found > 10 and < 11.50. A suppression is always preceded by an intercalation of *Āśvina*, *Kārttika* or *Mārgaśīrṣa* and may be followed by an intercalation of *Phālguna*. True intercalations of months are checked by fixing the moment of only one true new ☽. When after base > than indicated below the corresp. month has to be intercalated:	
	Caitra	M	32.523	>0.000	no intercalation		
	Vṛṣabha	T	61.288	2.085	*Caitra*		
	Vaiśākha	M	62.962	2.992	*Vaiśākha*		
	Mithuna	T	92.708	3.900	*Jyeṣṭha*		
	Jyeṣṭha	M	93.400	4.808	*Āṣāḍha*		
	Karka	T	124.353	5.715	*Śrāvaṇa*		
	Āṣāḍha	M	123.838	6.623	*Bhādrapada*		
	Siṃha	T	155.827	7.531	*Āśvina*		
	Śrāvaṇa	M	154.276	8.438	*Kārttika*		
	Kanyā	T	186.846	9.346	*Mārgaśīrṣa*		
	Bhādrapada	M	184.715	10.254	*Pauṣa*		
	Tulā	T	217.288	11.161	*Māgha*		
	Āśvina	M	215.153	12.069	*Phālguna*		
	Vṛścika	T	247.181	12.976	no intercalation		
	Kārttika	M	245.591	True	probable		
A	*Dhanus*	T	276.672	>0.00	*Caitra*	> 0.000	C
	Mārgaśīrṣa	M	276.029	0.20	*Vaiśākha*	> 30.353	
	Makara	T	305.990	1.60	*Jyeṣṭha*	> 61.287	
	Pauṣa	M	306.468	3.70	*Āṣāḍha*	> 92.708	
	Kumbha	T	335.438	5.90	*Śrāvaṇa*	> 124.352	
	Māgha	M	336.906	7.90	*Bhādrapada*	> 155.827	
	Mīna	T	365.259	9.40 !	*Āśvina*	> 186.846	
	Phālguna	M	367.344	10.25 !	*Kārttika*	> 217.288	
	Anomal. period ☉		365.259	(10.55)!	*Mārgaśīrṣa*	> 247.181	
D	Surplus of synodic period over			11.50	no intercalation or suppression of months		
	anomal. period ☽		1.976	28.90	*Caitra*, See Expl. § 20	> 0.000	

Mārgaśīrṣa is suppressed when 2 true new moons are found	:	< 247.181	> 276.672
Pauṣa ,, ,, ,, ,, ,, ,, ,, ,, ,,	:	< 276.672	> 305.990
Māgha ,, ,, ,, ,, ,, ,, ,, ,, ,,	:	< 305.990	> 335.437
Phālguna ,, intercalated ,, ,, ,, ,, ,, ,, ,,	:	> 335.438	< 365.259

	Julian months	leap year	common year
	January	0	0
	February	31	31
	March	60	59
	April	91	90
	May	121	120
	June	152	151
	July	182	181
E	August	213	212
	September	244	243
	October	274	273
	November	305	304
	December	335	334
	January	366	365
	February	397	396
	March	425	424
	April	456	455
	E		

An expired year of the *Kali Yuga* is reduced to a year A.D. by subtracting 3101

Gregorian Calendar.

The years 1700, 1800, 1900, 2100 etc. are common years.
Difference Greg.-Jul. year:
after Oct. 4, 1582: 10[d]
after Febr. 1700: 11[d]
,, ,, 1800: 12[d]
,, ,, 1900: 13[d]

of following year

425 in a leap year
456 in a leap year

F

Multiples of synodic period ☽		
The serial numbers between brackets to be used only when *Caitra* is an intercalated months and at the same time ☽ − ☉ at base > 28.9		
(1)		0.469 — 30
(2)	1	0.000
(3)	2	29.531
(4)	3	59.061
(5)	4	88.592
(6)	5	118.122
(7)	6	147.653
(8)	7	177.184
(9)	8	206.714
(10)	9	236.245
(11)	10	265.775
(12)	11	295.306
(13)	12	324.836
	13	354.367
G		

Multiples of anom. per. ☽	
27.55	
55.11	
82.66	
110.22	
137.77	
165.33	
192.88	
220.44	H
247.99	
275.55	
303.10	
330.66	
358.21	
385.76	
413.32	
440.87	
H	

TABLE III

B

No.	*tithis* (*śukla pakṣa*, bright half)	*karaṇas*		No.	*tithis* (*kṛṣṇa pakṣa*, dark half)	*karaṇas*	
1	*pratipadā*	*kiṃstughna*	0.000	1	*prathama*	*vālava*	14.765
		bava	0.492			*kaulava*	15.257
2	*dvitīyā*	*vālava*	0.984	2	*dvitīyā*	*taitila*	15.750
		kaulava	1.477			*gara*	16.242
3	*tṛitīyā*	*taitila*	1.969	3	*tṛitīyā*	*vaṇija*	16.734
		gara	2.461			*viṣṭi*	17.226
4	*caturthī*	*vaṇija*	2.953	4	*caturthī*	*bava*	17.718
		viṣṭi, bhadra	3.445			*vālava*	18.211
5	*pañcamī*	*bava*	3.937	5	*pañcamī*	*kaulava*	18.703
		vālava	4.430			*taitila*	19.195
6	*ṣaṣṭī*	*kaulava*	4.922	6	*ṣaṣṭī*	*gara*	19.687
		taitila	5.414			*vaṇija*	20.179
7	*saptamī*	*gara*	5.906	7	*saptamī*	*viṣṭi*	20.671
		vaṇija	6.398			*bava*	21.164
8	*aṣṭamī*	*viṣṭi*	6.890	8	*aṣṭamī*	*vālava*	21.656
		bava	7.383			*kaulava*	22.148
9	*navamī*	*vālava*	7.875	9	*navamī*	*taitila*	22.640
		kaulava	8.367			*gara*	23.132
10	*daśamī*	*taitila*	8.859	10	*daśamī*	*vaṇija*	23.624
		gara	9.351			*viṣṭi*	24.117
11	*ekādaśī*	*vaṇija*	9.844	11	*ekādaśī*	*bava*	24.609
		viṣṭi	10.336			*vālava*	25.101
12	*dvādaśī*	*bava*	10.828	12	*dvādaśī*	*kaulava*	25.593
		vālava	11.320			*taitila*	26.085
13	*trayodaśī*	*kaulava*	11.812	13	*trayodaśī*	*gara*	26.578
		taitila	12.304			*vaṇija*	27.070
14	*caturdaśī*	*gara*	12.797	14	*caturdaśī*	*viṣṭi*	27.562
		vaṇija	13.289			*śakuni*	28.054
15	*pūrṇimā*	*viṣṭi*	13.781	30	*amāvāsyā*	*nāga*	28.546
		bava	14.273			*catuṣpāda*	29.038

No mean expunction of a *tithi* is possible in a month beginning astr. $> 0^{d}469$ after Sunrise.
A mean *tithi* beginning $< 0^{d}016$ after sunrise is expunged.

A true *tithi* beginning $> 0^{d}103$ after sunrise cannot be expunged.
A true *tithi* beginning $< 0^{d}909$ after sunrise cannot be repeated.

B

A table of the serial numbers of the months in the lunisolar year.

	adhika = I, *nija* = II	*Caitra* I II	*Vaiś.* I II	*Jyeṣṭh.* I II	*Āṣāḍh* I II	*Śrāv.* I II	*Bhādr.* I II	*Āśvin.* I II	*Kārtt.* I II	*Mārg.* I II	*Pauṣa* I II	*Māgha* I II	*Phālg.* I II
	Common year	1	2	3	4	5	6	7	8	9	10	11	12
intercalated months	*Caitra*	1 2	3	4	5	6	7	8	9	10	11	12	13
	Vaiśākha	1	2 3	4	5	6	7	8	9	10	11	12	13
	Jyeṣṭha	1	2	3 4	5	6	7	8	9	10	11	12	13
	Āṣāḍha	1	2	3	4 5	6	7	8	9	10	11	12	13
	Śrāvaṇa	1	2	3	4	5 6	7	8	9	10	11	12	13
	Bāhdrap.	1	2	3	4	5	6 7	8	9	10	11	12	13
	Āśviṇa	1	2	3	4	5	6	7 8	9	10	11	12	13
	Kārttika	1	2	3	4	5	6	7	8 9	10	11	12	13
	Mārgaś.	1	2	3	4	5	6	7	8	9 10	11	12	13
	Pauṣa	1	2	3	4	5	6	7	8	9	10 11	12	13
	Māgha	1	2	3	4	5	6	7	8	9	10	11 12	13
	Phālguna	1	2	3	4	5	6	7	8	9	10	11	12 13
expunged	*Mārgaś.*									— —	10	11	12 (13)
	Pauṣa										— —	11	12 (13)
	Māgha											— —	12 (13)

A A

TABLE IV

A Table of the Equation of the Centre of the Moon for calculating *tithis*, etc.

	0. – 1		0. – 1		0. – 1		0.		0.		0.
0.0	000	5.0	624	10.0	686	13.8	002	19.0	385	24.0	300
1	990	1	620	1	692	9	012	1	388	1	294
2	981	2	616	2	698	14.0	021	2	391	2	287
3	971	3	613	3	705	1	031	3	394	3	280
4	962	4	610	4	711	2	040	4	397	4	273
5	952	5	607	5	718	3	050	5	399	5	266
6	943	6	604	6	725	4	059	6	402	6	259
7	934	7	601	7	732	5	068	7	404	7	252
8	924	8	599	8	739	6	078	8	406	8	244
9	915	9	597	9	747	7	087	9	408	9	236
1.0	906	6.0	595	11.0	754	8	096	20.0	409	25.0	229
1	897	1	593	1	762	9	106	1	410	1	221
2	887	2	591	2	770	15.0	115	2	411	2	212
3	878	3	590	3	778	1	124	3	412	3	204
4	869	4	589	4	786	2	133	4	413	4	196
5	860	5	588	5	794	3	142	5	413	5	188
6	852	6	587	6	802	4	150	6	414	6	179
7	843	7	587	7	810	5	159	7	414	7	171
8	834	8	586	8	819	6	168	8	414	8	162
9	825	9	586	9	827	7	176	9	413	9	153
2.0	817	7.0	586	12.0	836	8	185	21.0	413	26.0	144
1	809	1	587	1	845	9	193	1	412	1	136
2	800	2	587	2	854	16.0	202	2	411	2	127
3	792	3	588	3	862	1	210	3	410	3	118
4	784	4	589	4	871	2	218	4	408	4	108
5	776	5	590	5	880	3	226	5	406	5	099
6	768	6	592	6	889	4	234	6	404	6	090
7	760	7	593	7	899	5	241	7	402	7	081
8	753	8	595	8	908	6	249	8	400	8	071
9	745	9	597	9	917	7	257	9	398	9	062
3.0	738	8.0	599	13.0	926	8	264	22.0	395	27.0	053
1	731	1	602	1	936	9	271	1	392	1	043
2	723	2	604	2	945	17.0	278	2	389	2	034
3	717	3	607	3	955	1	285	3	386	3	024
4	710	4	610	4	964	2	292	4	382	4	015
5	703	5	614	5	974	3	298	5	378	5	005
6	697	6	617	6	983	4	305	6	374		0.–1
7	690	7	621	7	993	5	311	7	370	6	996
8	684	8	625	13.78	000	6	317	8	366	7	986
9	678	9	629			7	323	9	362	8	977
4.0	672	9.0	633			8	329	23.0	357	9	967
1	667	1	637			9	335	1	352	28.0	958
2	661	2	642			18.0	340	2	347	1	948
3	656	3	647			1	345	3	342	2	939
4	651	4	652			2	350	4	336	3	929
5	646	5	657			3	355	5	331	4	920
6	641	6	662			4	360	6	325	5	911
7	636	7	668			5	365	7	319	6	902
8	632	8	674			6	369	8	313	7	892
9	628	9	680			7	373	9	307	8	883
5.0	624	10.0	686			8	377	24.0	300	9	874
1	620	1	692			9	381	1	294	29.0	865
2	616	2	698			19.0	385	2	287	1	856

9	
0	0
1	1
2	2
3	3
4	4
5	5
7	6
8	7
9	8
8	
0	0
1	1
2	2
4	3
5	4
6	5
7	6
9	7
7	
0	0
1	1
3	2
4	3
5	4
7	5
9	6
6	
0	0
1	1
3	2
5	3
6	4
8	5
5	
0	0
1	1
3	2
5	3
7	4
9	5
4	
0	0
2	1
4	2
7	3
9	4
3	
0	0
2	1
5	2
9	3
2	
0	0
4	1
8	2

Long. East fr. Gr.	Correction for terrestrial longitude 0, (A)
65	—0 30
70	—0 16
71	—0 13
72	—0 10
73	—0 08
74	—0 05
75	—0 02
76	+0 01
77	0 03
78	0 06
79	0 09
80	0 12
81	0 15
82	0 17
83	0 20
84	0 23
85	0 26
90	0 40

d resp. d+△	equation of time (B) 3000, argument d, 0,	equation of time (B) 5000, argument d, 0,		Sunrise in apparent time (D) $\varphi = 10°$, argument d + △, 0,	$\varphi = 20°$ 0,	$\varphi = 22°$ 0,	$\varphi = 24°$ 0,	$\varphi = 26°$ 0,
0	—0 05	—0 06	—	0 08	0 21	0 23	0 26	0 29
10	— 8	— 8	—	6	19	21	24	27
20	— 10	— 10	—	5	17	18	20	23
30	— 12	— 10	—	3	13	15	16	18
40	— 12	— 9	—	1	10	11	13	14
50	— 11	— 8	—	0	6	7	8	10
60	— 10	— 6	—	0	2	3	3	4
70	— 8	— 4	+	2	2	2	2	2
80	— 5	— 2		4	5	6	6	6
90	— 3	—0 00		6	10	10	11	12
100	0 00	+0 01		8	14	15	16	17
110	+ 2	2		10	17	18	20	22
120	4	3		12	20	22	24	26
130	5	2		13	23	25	28	31
140	5	0 00		14	25	28	31	34
150	5	0 00		15	27	30	33	36
160	4	—0 01		15	28	31	34	37
170	3	— 3		14	28	31	34	37
180	1	— 4		13	27	30	33	36
190	0 00	— 5		13	25	28	30	33
200	— 1	— 5		12	23	25	28	30
210	— 2	— 4		10	20	22	24	26
220	— 2	— 2		8	17	19	20	22
230	— 1	0 00		6	14	15	16	17
240	0 00	+ 2		4	10	10	11	12
250	+ 2	5		2	6	6	7	7
260	4	7		1	2	2	2	2
270	5	9	—	1	2	3	3	4
280	8	10	—	3	6	7	8	9
290	9	11	—	4	10	11	12	14
300	10	11	—	6	13	15	17	18
310	9	10	—	8	16	18	20	23
320	8	8	—	9	19	21	23	26
330	6	5	—	10	21	23	25	28
340	4	2	—	10	22	25	28	30
350	0 00	—0 01	—	10	23	25	28	31
360	— 1	— 5	—	9	22	24	27	30
370	— 7	— 7	—	8	21	23	26	28

K.Y. exp. (C)	△
2200	—16
23	—15
24	—15
25	—14
26	—13
27	—12
28	—11
29	—11
30	—10
31	— 9
32	— 8
33	— 8
34	— 7
35	— 6
36	— 5
37	— 5
38	— 4
39	— 3
40	— 2
41	— 2
42	— 1
43	0
44	0
45	+ 1
46	+ 2
47	+ 3
48	+ 3
49	+ 4
50	+ 5
51	+ 6
52	+ 6

A Table for converting decimals of the day into *ghaṭikās* and *palas*. E.g. 0.769 = 45 g 36 p + 32 = 46 g 8 p (E)

Second decimal:	0	1	2	3	4	5	6	7	8	9
first decimal	gh. p	gh. p	gh. p	gh. p	gh. p	gh. p	gh. p	gh. p	gh. p	gh. p
0	0 0	0 36	1 12	1 48	2 24	3 0	3 36	4 12	4 48	5 24
1	6 0	6 36	7 12	7 48	8 24	9 0	9 36	10 12	10 48	11 24
2	12 0	12 36	13 12	13 48	14 24	15 0	15 26	16 12	16 48	17 24
3	18 0	18 36	19 12	19 48	20 24	21 0	21 36	22 12	22 48	23 24
4	24 0	24 36	25 12	25 48	26 24	27 0	27 36	28 12	28 48	29 24
5	30 0	30 36	31 12	31 48	32 24	33 0	33 36	34 12	34 48	35 24
6	36 0	36 36	37 12	37 48	38 24	39 0	39 36	40 12	40 48	41 24
7	42 0	42 36	43 12	43 48	44 24	45 0	45 36	46 12	46 48	47 24
8	48 0	48 36	49 12	49 48	50 24	51 0	51 36	52 12	52 48	53 24
9	54 0	54 36	55 12	55 48	56 24	57 0	57 36	58 12	58 48	59 24
third decimal:	0 0	0 4	0 7	0 11	0 14	0 18	0 22	0 25	0 29	0 32

SECOND AUXILIARY TABLE

MIX
Papier aus verantwortungsvollen Quellen
Paper from responsible sources
FSC® C105338

If you have any concerns about our products,
you can contact us on
ProductSafety@springernature.com

In case Publisher is established outside the EU,
the EU authorized representative is:
Springer Nature Customer Service Center GmbH
Europaplatz 3, 69115 Heidelberg, Germany

Printed by Libri Plureos GmbH
in Hamburg, Germany